FOUNDATIONS OF SOFTWARE TESTING

ISTQB CERTIFICATION

FIFTH EDITION

Erik van Veenendaal

Rex Black

Dorothy Graham

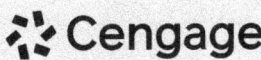
Cengage

Australia • Brazil • Canada • Mexico • Singapore • United Kingdom • United States

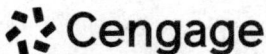

***Foundations of Software Testing: ISTQB Certification,* Fifth Edition**
Erik van Veenendaal, Rex Black, Dorothy Graham

Publisher: Annabel Ainscow

List Manager: Virginia Thorp

Marketing Manager: Hugh Callaghan

Content Project Manager:
  Narmada Kaushal

Manufacturing Buyer: Eyvett Davis

Manufacturing Manager: Elaine Bevan

Typesetter: Lumina Datamatics Ltd.

Text Design: Lumina Datamatics Ltd.

Cover Design: Cyan Designs

Cover Image(s):
  ©Clau3Dia\shutterstock

For product information and technology assistance, contact us at
emea.info@cengage.com

For permission to use material from this text or product and for permission queries, email **emea.permissions@cengage.com**

British Library Cataloguing-in-Publication Data

A catalogue record for this book is available from the British Library.

ISBN: 978-1-4737-9588-4

Cengage Learning, EMEA
Cheriton House, North Way
Andover, Hampshire, SP10 5BE
United Kingdom

Cengage Learning is a leading provider of customized learning solutions with employees residing in nearly 40 different countries and sales in more than 125 countries around the world. Find your local representative at: **cengage.uk**

To learn more about Cengage platforms and services, register or access your online learning solution, or purchase materials for your course, visit **cengage.uk**

Printed in United Kingdom from CPI Antony Rowe
Print Number: 01   Print Year: 2025

CONTENTS

FIGURES AND TABLES

ACKNOWLEDGEMENTS

The materials in this book are based on the ISTQB Foundation Syllabus v 4.0 (2023). The Foundation Syllabus is copyrighted to the ISTQB (International Software Testing Qualification Board). Permission has been granted by the ISTQB to the authors to use these materials as the basis of a book, provided that recognition of authorship and copyright of the Syllabus itself is given.

The ISTQB Glossary of Testing Terms, released as version 4.2 by the ISTQB in 2023, is used as the source of definitions in this book.

Be aware that there are some defects in this book! The Syllabus, Glossary and this book were written by people – and people make mistakes. Just as with testing, we have applied reviews and tried to identify as many defects as we could, but we also needed to release the manuscript to the publisher. Please let us know of defects that you find in our book so that we can correct them in future printings.

The authors wish to acknowledge the contribution of Isabel Evans to the first edition of this book. We also acknowledge contributions to this edition from Rogier Ammerlaan, Katalin Balla, Mark Fewster, Claire Horgan, Kari Kakkonen, Tal Pe'er and Adriana Chirstina Urse.

Erik van Veenendaal, The Netherlands
Dorothy Graham, Macclesfield, UK
Rex Black, Texas, USA
2024

PREFACE

The purpose of this book is to support the ISTQB Foundation Syllabus v 4.0 (2023), which is the basis for the International Foundation Certificate in Software Testing (CTFL). The authors have been involved in helping to establish this qualification, donating their time and energy to the Syllabus, terminology Glossary and the International Software Testing Qualifications Board (ISTQB).

The authors of this book are all passionate about software testing. All have been involved in this area for most or all of their working lives, and have contributed to the field through practical work, training courses and books. They have written this book to help promote the discipline of software testing.

The initial idea for this collaboration came from Erik van Veenendaal, author of *The Testing Practitioner*, a book to support the ISEB Software Testing Practitioner Certificate. The other authors agreed to work together as equals on this book. Please note that the order of the authors' names does not indicate any seniority of authorship, but simply which author was the last to update the book as the Foundation Syllabus evolved.

We intend that this book will increase your chances of passing the Foundation Certificate exam. If you are taking a course (or class) to prepare for the exam, this book will give you detailed and additional background about the topics you have covered. If you are studying for the exam on your own, this book will help you be more prepared. This book will give you information about the topics covered in the Syllabus, as well as worked exercises and practice exam questions (including a full 40-question mock exam paper in Chapter 7).

This book goes beyond the Syllabus in a number of areas and is a useful reference work about software testing in general. In some places in the book, we note additional material as an optional extra section. If you are focusing solely on the exam, then you can skip these sections (and come back to them later as they are useful). The Foundation Certificate represents a distilling of the essential aspects of software testing at the time of writing (2024), and even if you are not interested in taking the exam, this book will give you a good grounding in software testing.

ISTQB AND CERTIFICATION

ISTQB stands for International Software Testing Qualifications Board and is an organization consisting of software testing professionals from each of the countries who are members of the ISTQB. Each representative is a member of a Software Testing Board in their own country. The purpose of the ISTQB is to provide internationally accepted and consistent qualifications in software testing. ISTQB sets the Syllabus and gives guidelines for each member country to implement the qualification in their own country. The Foundation Certificate is the first internationally accepted qualification in software testing and its Syllabus forms the basis of this book.

From the first qualification in 1998 until June 2023, 1.2 million people have taken the Foundation Certificate exam administered by a National Board of the ISTQB, or by an Exam Board contracted to a National Board. A total of 914,000 certifications have been issued, in over 130 countries. The Certified Tester Foundation Level (CTFL) is by far the most popular of the ISTQB certifications. All ISTQB Member Boards and Exam Providers recognize each other's Foundation Certificates as valid.

The ISTQB qualification is independent of any individual training provider. Any training organization can offer a course based on this publicly available Syllabus. However, the Member Boards associated with ISTQB give special approval to organizations that meet their requirements for the quality of the training. Such organizations are accredited and are allowed to have an invigilator or proctor from an authorized Member Board or Exam Provider to give the exam as part of an accredited in-person course. The exam is also available independently from accrediting organizations or Member Boards, as well as online.

Why is certification of testers important? The objectives of the qualification include the following:

- Recognition for testing as an essential and professional software engineering specialization.
- Enabling professionally qualified testers to be recognized by employers, customers and peers.
- Raising the profile of testers.
- Promoting consistent and good testing practices within all software engineering disciplines internationally, for reasons of opportunity, communication and sharing of knowledge and resources internationally.

FINDING YOUR WAY AROUND THIS BOOK

This book is divided into seven chapters. The first six chapters of the book each cover one chapter of the Syllabus, and each has some practice exam questions.

Chapter 1 is the start of understanding. We'll look at some fundamental questions: what is testing and why is it necessary? We'll examine why testing is not just running tests, and consider fundamental principles of software testing. We'll also look at testware, test activities, test roles, essential skills and good practices in testing.

In Chapter 2, we'll concentrate on testing in relation to common software development models, including sequential and iterative models, including Agile development. We'll look at test-first approaches such as shift-left, and look at DevOps and retrospectives in testing. We will distinguish different levels of testing and different types of testing which are used at different stages in the software development life cycle, and consider maintenance testing.

In Chapter 3, we'll concentrate on test techniques that can be used early in the software development life cycle. These include reviews and static analysis: tests done before compiling the code. We consider early and frequent stakeholder feedback in a review process, and when different review types are most beneficial.

Chapter 4 covers test techniques. We'll show you black-box techniques including equivalence partitioning, boundary value analysis, decision tables and state transition testing. We look at white-box testing including statement and branch coverage. We cover experience-based techniques, as well as collaboration-based approaches. This chapter is about how to become a better tester in terms of designing tests. There are exercises for the techniques included in this chapter.

Chapter 5 is about the management and control of testing and tests. We cover test planning, including estimation, entry and exit criteria, prioritizing, the test pyramid and test quadrants. We look at risk management: risk levels, risk assessment, product and project risks and risk mitigation. We look at test monitoring, test control and test completion. Finally, we look at managing the tests themselves: configuration management and defect management. Writing a good defect report is a key skill for a good tester, so we have an exercise for that too.

In Chapter 6, we'll show you how tools support all the activities in the test process, and talk about the risks and benefits of test automation.

Chapter 7 contains general advice about taking the exam and has a full 40-question mock paper which conforms to the requirements of a real ISTQB exam. This is a key learning aid to help you pass the real exam.

The appendices of the book include a full list of references and a copy of the ISTQB testing terminology Glossary, as well as the answers to all the practice exam questions.

TO HELP YOU USE THE BOOK

1 **Get a copy of the Syllabus:** You should download the Syllabus from the ISTQB website so that you have the current version, and so that you can check off the Syllabus objectives as you learn. This is available to download at the ISTQB website.

2 **Understand what is meant by learning objectives and knowledge levels:** In the Syllabus, you will see learning objectives and knowledge (or cognitive) levels at the start of each section of each chapter. These indicate what you need to know and the depth of knowledge required for the exam. We have used the timings in the Syllabus and knowledge levels to guide the space allocated in the book, both for the text and for the exercises. You will see the learning objectives and knowledge levels at the start of each section within each chapter. The knowledge levels expected by the Syllabus are:

- **K1: remember, recognize, recall:** you will recognize, remember and recall a term or concept. For example, you could recognize the definition of failure as 'An event in which a component or system does not meet its requirements within specified limits'.
- **K2: explain, give reasons, compare, classify, summarize:** you can select the reasons or explanations for statements related to the topic, and can summarize, compare, classify and give examples for the testing concept. For example, you could explain that one reason why tests should be designed as early as possible is to find defects when they are cheaper to remove.
- **K3: apply:** you can select the correct application of a concept or technique and apply it to a given context. For example, you could identify boundary values for valid and invalid partitions, and you could select test cases from a given state transition diagram in order to cover all valid transitions.

Remember, as you go through the book, if a topic has a learning objective marked K1 you just need to recognize it. If it has a learning objective of K3 you will be expected to apply your knowledge in the exam, for example.

3 **Use the Glossary of terms:** Each chapter of the Syllabus has a number of terms listed in it. You are expected to remember these terms at least at K1 level, even if they are not explicitly mentioned in the learning objectives. You will see a number of **definitions** throughout this book, as in the sidebar.

Definition A description of the meaning of a word.

All definitions of software testing terms (called keywords in the chapters) are taken from the *ISTQB Glossary* (version 4.2). The easiest way to find a definition is to search in the ISTQB online glossary at glossary.istqb.org. A copy of the Glossary (2024) is also at the back of the book. All the terms that are specifically mentioned in the Syllabus, that is, the ones you need to learn for the exam, are mentioned in each section of this book. Note that sometimes Glossary terms are updated, so the latest definition will be online.

You will notice that some terms in the Glossary at the back of this book are underlined. These are terms that are mentioned specifically as keywords in the Syllabus. These are the terms that you need to be familiar with for the exam. These terms are also shown in **bold** the first time they appear in this book.

4 **Use the references sensibly:** We have referenced all the books used by the Syllabus authors when they constructed the Syllabus. You will see these underlined in the list at the end of the book. We also added references to some other books, papers and websites that we thought useful or which we referred to when writing. You do not need to read all referenced books for the exam! However, you may find some of them useful for further reading to increase your knowledge after the exam, and to help you apply some of the ideas you will come across in this book.

5 **Do the practice exams:** When you get to the end of a chapter (for Chapters 1 to 6), answer the exam questions, and then turn to 'Answers to the Sample Exam Questions' to check if your answers were correct. After you have completed all of the six chapters, then take the full mock exam in Chapter 7. If you would like the most realistic exam conditions, then allow yourself just an hour to take the exam in Chapter 7. Also take the free sample exams from the ISTQB website. You can download both the exam and the answers including justifications for the correct (and wrong) answers.

CHAPTER ONE
Fundamentals of testing

In this chapter, we will introduce you to the fundamentals of testing: what software testing is and why testing is needed, including its limitations, objectives and purpose; the principles behind testing; the process that testers follow, including activities, tasks and work products; and some of the psychological factors that testers must consider in their work. By reading this chapter you will gain an understanding of the fundamentals of testing and be able to describe those fundamentals.

Note that the learning objectives start with 'FL' rather than 'LO' to show that they are learning objectives for the Foundation Level qualification.

1.1 WHAT IS TESTING?

> **SYLLABUS LEARNING OBJECTIVES FOR 1.1 WHAT IS TESTING? (K2)**
>
> **FL-1.1.1** Identify typical test objectives (K1)
>
> **FL-1.1.2** Differentiate testing from debugging (K2)

In this section, we will kick off the book by looking at what testing is, some misconceptions about testing, the typical objectives of testing and the difference between testing and debugging.

Within each section of this book, there are terms that are important—they are used in the section (and may be used elsewhere as well). They are listed in the Syllabus as keywords, which means that you need to know the definition of the term and it could appear in an exam question. We will give the definition of the relevant keyword terms in the margin of the text, and they can also be found in the Glossary (including the ISTQB online Glossary at https://glossary.istqb.org). We also show the keyword in **bold** within the section or subsection where it is defined and discussed.

In this section, the relevant keyword terms are **debugging**, **test object**, **test objective**, **testing**, **validation** and **verification**.

Software is everywhere

The last 100 years have seen an amazing human triumph of technology. Diseases that once killed and paralyzed are routinely treated or prevented—or even eradicated entirely, as with smallpox. Some children who stood amazed as they watched the first gasoline-powered automobile in their town are alive today, having seen people walk on the moon, an event that happened before a large percentage of today's workforce was even born.

Perhaps the most dramatic advances in technology have occurred in the arena of information technology. Software systems, in the sense that we know them, are a recent innovation, less than a century old, but have already transformed daily life around the world. Thomas Watson, the one-time head of IBM, famously predicted that only about five computers would be needed in the whole world. This vastly inaccurate prediction was based on the idea that information technology was useful only for business and government applications, such as banking, insurance and conducting a census. (The Hollerith punch-cards used by computers at the time Watson made his prediction were developed for the US census.) Now, everyone who drives a car is using a machine not only designed with the help of computers, but which also contains more computing power than the computers used by NASA to get Apollo missions to and from the moon. Mobile phones are now essentially handheld computers that get smarter with every new model. The Internet of Things (IoT) now gives us the ability to see who is at our door or turn on the lights when we are nowhere near our home.

However, in the software world, the technological triumph has not been perfect. Almost every living person has been touched by information technology, and most of us have dealt with the frustration and wasted time that occurs when software fails and exhibits unexpected behaviours. Some unfortunate individuals and companies have experienced financial loss or damage to their personal or business reputations as a result of defective software. A highly unlucky few have even been injured or killed by software failures, including by self-driving cars.

One way to help overcome such problems is software testing, when it is done well. Testing covers activities throughout the life cycle and can have a number of different objectives, as we will see in Section 1.1.1.

Testing is more than running tests

Testing The process within the software development lifecycle that evaluates the quality of a component or system and related work products.

An ongoing misperception about **testing**, although less common these days, is that it only involves running tests. Specifically, some people think that testing involves nothing beyond carrying out some sequence of actions on the system under test, submitting various inputs along the way and evaluating the observed results. Certainly, these activities are one element of testing—specifically, these activities make up the bulk of the test execution activities—but there are many other activities involved in the test process.

We will discuss the test process in more detail in Section 1.4, but testing also includes (in addition to test execution): test planning, analyzing, designing and implementing tests, reporting test progress and results, and reporting defects. As you can see, there is a lot more to it than just running tests.

Notice that there are major test activities both before and after test execution. In addition, you will see that testing includes both static and dynamic testing. Static testing is any evaluation of the software or related work products (such as requirements specifications or user stories) that occurs without executing the software itself. Dynamic testing is an evaluation of that software or related work products that does involve executing the software. Testing not only includes a number of pre-execution and post-execution activities that non-testers often do not consider 'testing' but also includes software quality activities (for example, requirements reviews and static analysis of code) that non-testers (and even sometimes testers) often do not consider 'testing' either.

The reason for this broad definition is that both dynamic testing (at whatever level) and static testing (of whatever type) often enable the achievement of similar project objectives. Dynamic testing and static testing also generate information that can help achieve an important process objective—that of understanding and improving the software development and testing processes. Dynamic testing and static testing are complementary activities, each able to generate information that the other cannot.

Testing is more than verification

Another common misconception about testing is that it is only about checking correctness, that is, that the system corresponds to its requirements, user stories or other specifications. Checking against a specification (called **verification**) is certainly part of testing, where we are asking the question, 'Have we built the system correctly?' Note the emphasis in the definition on 'specified requirements'.

But just conforming to a specification is not sufficient testing, as we will see in Section 1.3 (Principle 7 Absence-of-defects is a fallacy). We also need to test to see if the delivered software and system will meet user and stakeholder needs and expectations in its operational environment. Often it is the tester who becomes the advocate for the end-user in this kind of testing, which is called **validation**. Here we are asking the question, 'Have we built the right system?'

In every development life cycle, a part of testing is focused on verification testing and a part is focused on validation testing. Verification is concerned with evaluating a work product, component or system to determine whether it meets the requirements set. In fact, verification focuses on the question, 'Is the deliverable built according to the specification?' Validation is concerned with evaluating a work product, component or system to determine whether it meets the user and stakeholder needs and requirements. Validation focuses on the question, 'Is the deliverable fit for purpose, for example does it provide a solution to the problem?'

> **Verification**
> Confirmation by examination and through provision of objective evidence that specified requirements have been fulfilled.
>
> **Validation**
> Confirmation by examination that a work product matches a stakeholder's needs.

1.1.1 Typical objectives of testing

The following are some **test objectives** given in the Foundation Syllabus:

> **Test objective** The purpose for testing.

- To evaluate work products such as requirements, user stories, design and code including the use of static testing techniques, such as reviews.

- To verify whether all specified requirements have been fulfilled, for example in the resulting system.

- To validate whether the test object is complete and works as the users and other stakeholders expect, for example together with user or stakeholder groups.

- To build confidence in the level of quality of the test object, such as when those tests considered highest risk pass, and when the failures that are observed in the other tests are considered acceptable.

- To trigger failures and find defects; this is typically a prime focus for software testing.

- To provide sufficient information to stakeholders to allow them to make informed decisions, especially regarding the level of quality of the **test object**, for example by the satisfaction of entry or exit criteria.

> **Test object** The work product to be tested.

- To ensure the required level of coverage of a test object is achieved; the level of coverage to be achieved is typically one of the exit criteria for testing.

- To reduce the level of risk of inadequate software quality, for example previously undetected failures occurring in operation.
- To comply with contractual, legal or regulatory requirements or standards and/ or to verify the test object's compliance with such requirements or standards.

Another good objective for testing is to help to prevent defects, such as when early test activities, for example requirements reviews or early test design, identify defects in requirements specifications that are removed before they cause defects in the design specifications and subsequently the code itself. Both reviews and test design serve as a verification and validation of these test basis documents that will reveal problems that otherwise would not surface until test execution, potentially much later in the project.

These objectives are not universal. Different test viewpoints, test levels and test stakeholders can have different objectives. While many levels of testing, such as component, integration and system testing, focus on discovering as many failures as possible in order to find and remove defects, in acceptance testing the main objective is confirmation of correct system operation (at least under normal conditions), together with building confidence that the system meets its requirements. The context of the test object and the software development life cycle will also affect what test objectives are appropriate. Let's look at some examples to illustrate this.

When evaluating a software package that might be purchased or integrated into a larger software system, the main objective of testing might be the assessment of the quality of the software. Defects found may not be fixed, but rather might support a conclusion that the software be rejected.

During component testing, one objective at this level may be to achieve a given level of code coverage by the component tests, that is, to assess how much of the code has actually been exercised by a set of tests and to add additional tests to exercise parts of the code that have not yet been covered/tested. Another objective may be to find as many failures as possible so that the underlying defects are identified and fixed as early as possible.

During user acceptance testing, one objective may be to confirm that the system works as expected (validation) and satisfies requirements (verification). Another objective of testing here is to focus on providing stakeholders with an evaluation of the risk of releasing the system at a given time. Evaluating risk can be part of a mix of objectives, or it can be an objective of a separate level of testing, as when testing a safety-critical system, for example.

During maintenance testing, our objectives often include checking whether developers have introduced any regressions (new defects not present in the previous version) while making changes. Some forms of testing, such as operational testing, focus on assessing quality characteristics such as reliability, security, performance or availability.

1.1.2 Testing and debugging

Debugging The process of finding, analyzing and removing the causes of failures in a component or system.

Let's end this section by saying what testing is not, but is often thought to be. Testing is not **debugging**. While dynamic testing often locates failures which are caused by defects, and static testing often locates defects themselves, testing does not fix defects. It is during debugging, a development activity, that a member of the project team finds, analyzes and removes the defect, the underlying cause of the failure. After debugging, there is a further testing activity associated with the defect, which is called confirmation testing. This activity ensures that the fix does indeed resolve the failure.

In terms of roles, dynamic testing is a testing role, debugging is a development role and confirmation testing is again a testing role. However, in Agile teams, this distinction may be blurred, as testers may be involved in debugging and component testing.

Further information about software testing concepts can be found in the ISO standard ISO/IEC/IEEE 29119-1 [2022].

1.2 WHY IS TESTING NECESSARY?

SYLLABUS LEARNING OBJECTIVES FOR 1.2 WHY IS TESTING NECESSARY? (K2)

FL-1.2.1   **Exemplify why testing is necessary (K2)**

FL-1.2.2   **Recall the relationship between testing and quality assurance (K1)**

FL-1.2.3   **Distinguish between root cause, error, defect and failure (K2)**

In this section, we discuss how testing contributes to success and the relationship between testing and quality assurance. We will describe the difference between errors, defects and failures and illustrate how software defects or bugs can cause problems for people, the environment or a company. We will draw important distinctions between defects, their root causes and their effects.

As we go through this section, watch for the Syllabus terms **defect**, **error**, **failure**, **quality**, **quality assurance** and **root cause**.

Testing can help to reduce the risk of failures occurring during operation, provided it is carried out in a rigorous way, including reviews of documents and other work products. Testing both verifies that a system is correctly built and validates that it will meet users' and stakeholders' needs, even though no testing is ever exhaustive (refer to Principle 2 in Section 1.3, Exhaustive testing is impossible). In some situations, testing may not only be helpful but may also be necessary to meet contractual or legal requirements or to conform to industry-specific standards, such as automotive or safety-critical systems.

1.2.1 Testing's contributions to success

As we mentioned in Section 1.1, we have all experienced software problems, for example an app fails in the middle of doing something, a website freezes while taking your payment (did it go through or not?) or inconsistent prices for exactly the same flights on travel sites. Failures like these are annoying, but failures in safety-critical software can be life-threatening, such as in medical devices or self-driving cars.

The use of appropriate test techniques, applied with the right level of test expertise at the appropriate test levels and points in the software development life cycle, can be of significant help in identifying problems so that they can be fixed before the software or system is released into use. Here are some examples where testing could contribute to more successful systems:

- Testers often become the representatives of the users throughout the software development lifecycle (SDLC). Having testers involved in requirements reviews or user story refinement could detect defects in these work products before any design or coding is done for the functionality described. Identifying and removing defects at this stage reduces the risk of the wrong software (incorrect or untestable) being developed.

- Having testers work closely with system designers and developers while the system is being designed can increase each party's understanding of the design and how to test it. Since misunderstandings are often the cause of defects in software, having a better understanding at this stage can reduce the risk of design defects. A bonus is that tests can be identified from the design—thinking about how to test the system at this stage often results in better design. Having testers work closely with developers while the code is under development can increase each party's understanding of the code and how to test it. As with design, this increased understanding, and the knowledge of how the code will be tested, can reduce the risk of defects in the code (and in the tests).

- Having testers verify and validate the software prior to release can detect failures that might otherwise have been missed—this is traditionally where the focus of testing has been. Having defects identified early means that both finding and fixing them is more cost-effective. As we see with the previous examples, if we leave it until release, we will not be nearly as efficient as we would have been if we had caught these defects earlier. However, it is still necessary to test just before release, and testers can also help to support debugging activities, for example by running confirmation and regression tests. Thus, testing can help the software meet stakeholder needs and satisfy requirements.

- Testers may also be required by regulatory bodies to meet contractual or legal requirements and conform to standards, such as for safety-critical systems.

In addition to these examples, achieving the defined test objectives (refer to Section 1.1.1) also contributes to the overall success of software development and maintenance.

1.2.2 Quality assurance and testing

Quality assurance
Activities focused on providing confidence that quality requirements will be fulfilled.

Is **quality assurance** (QA) the same as testing? Many people refer to 'doing QA' when they are actually testing, and some job titles refer to QA when they really mean testing. The two are not the same. QA is actually one part of a larger concept, quality management, which refers to all activities that direct and control an organization with regard to quality in all aspects. Quality affects not only software development but also human resources (HR) procedures, delivery processes and even the way people answer the company's telephones.

Quality management consists of a number of activities, including QA and quality control (as well as setting quality objectives, quality planning and quality improvement). QA is associated with ensuring that a company's standard ways of performing various tasks are carried out correctly. Such procedures may be written in a quality handbook that everyone is supposed to follow. The idea is that if processes are carried out correctly, then the products produced will be of higher **quality**. Root cause analysis and retrospectives are used to help to improve processes for more effective QA. If they are following a recognized quality management standard, companies may be audited to ensure that they do actually follow their prescribed processes (say what you do, and do what you say).

Quality control is concerned with the quality of products rather than processes, to ensure that they have achieved the desired level of quality. Testing is looking at work products, including software, so it is actually a quality control activity rather than a QA activity, despite common usage. However, testing also has processes that should be followed correctly, so QA does support good testing in this way. Sections 1.1.1 and 1.2.1 describe how testing contributes to the achievement of quality.

So, we see that testing plays an essential supporting role in delivering quality software. However, testing by itself is not sufficient. Testing should be integrated into a complete, team-wide and development process-wide set of activities for QA. Proper application of standards, training of staff, the use of retrospectives to learn lessons from defects and other important elements of previous projects, rigorous and appropriate software testing: all of these activities and more should be deployed by organizations to ensure acceptable levels of quality and quality risk upon release.

> **Quality** The degree to which a work product satisfies stated and implied needs of its stakeholders.

1.2.3 Errors, defects and failures

Why does software fail? Part of the problem is that, ironically, while computerization has allowed dramatic automation of many professions, software engineering remains a human-intensive activity. And humans are fallible beings. So, software is fallible because humans are fallible.

The precise chain of events goes something like this. A developer makes an **error** (or mistake), such as forgetting about the possibility of inputting an excessively long string into a field on a screen. The developer thus puts a **defect** (or fault or bug) into the program, such as omitting a check on input strings for length prior to processing them. When the program is executed, if the right conditions exist (or the wrong conditions, depending on how you look at it), the defect may result in unexpected behaviour, that is, the system exhibits a **failure**, such as accepting an over-long input that it should reject, with subsequent corruption of other data.

Other sequences of events can result in eventual failures, too. A business analyst can introduce a defect into a requirement, which can escape into the design of the system and further escape into the code. For example, a business analyst might say that an e-commerce system should support 100 simultaneous users, but actually peak load should be 1,000 users. If that defect is not detected in a requirements review (refer to Chapter 3), it could escape from the requirements phase into the design and implementation of the system. Once the load exceeds 100 users, resource utilization may eventually spike to dangerous levels, leading to reduced response time and reliability problems.

A technical writer can introduce a defect into the online help screens. For example, suppose that an accounting system is supposed to multiply two numbers together, but the help screens say that the two numbers should be added. In some cases, the system will appear to work properly, such as when the two numbers are both 0 or both 2. However, most frequently the program will exhibit unexpected results (at least based on the help screens).

> **Error** (mistake) A human action that produces an incorrect result.
>
> **Defect** (bug, fault) An imperfection or deficiency in a work product where it does not meet its requirements or specifications.
>
> **Failure** An event in which a component or system does not perform a required function within specified limits.

Sometimes a failure occurs due to data issues, such as the UK Air Traffic Control 'glitch' in August 2023, where 1,500 flights were cancelled and more were delayed over several days. This was due to a flight plan containing two waypoints outside the UK with the same name, but which were not the same location, and the flight plan had not specified an exit point from UK airspace. Refer to [NATS 2023].

So, human beings are fallible and thus, when they work, they sometimes introduce defects. It is important to point out that the introduction of defects is not a purely random accident, though some defects may be introduced randomly, such as when a phone rings and distracts a systems engineer in the middle of a complex series of design decisions. The rate at which people make errors increases when they are under time pressure, when they are working with complex systems, interfaces or code, and when they are dealing with changing technologies or highly interconnected systems.

While we commonly think of failures being the result of 'bugs in the code', a significant number of defects are introduced in work products such as requirements specifications and design specifications. Capers Jones reports that about 20% of defects are introduced in requirements and about 25% in design. The remaining 55% are introduced during implementation or repair of the code, metadata or documentation [Jones 2008]. Other experts and researchers have reached similar conclusions, with one organization finding that as many as 75% of defects originate in requirements and design. Figure 1.1 shows four typical scenarios, the upper stream being correct requirements, design and implementation, the lower three streams showing defect introduction at some phase in the software life cycle.

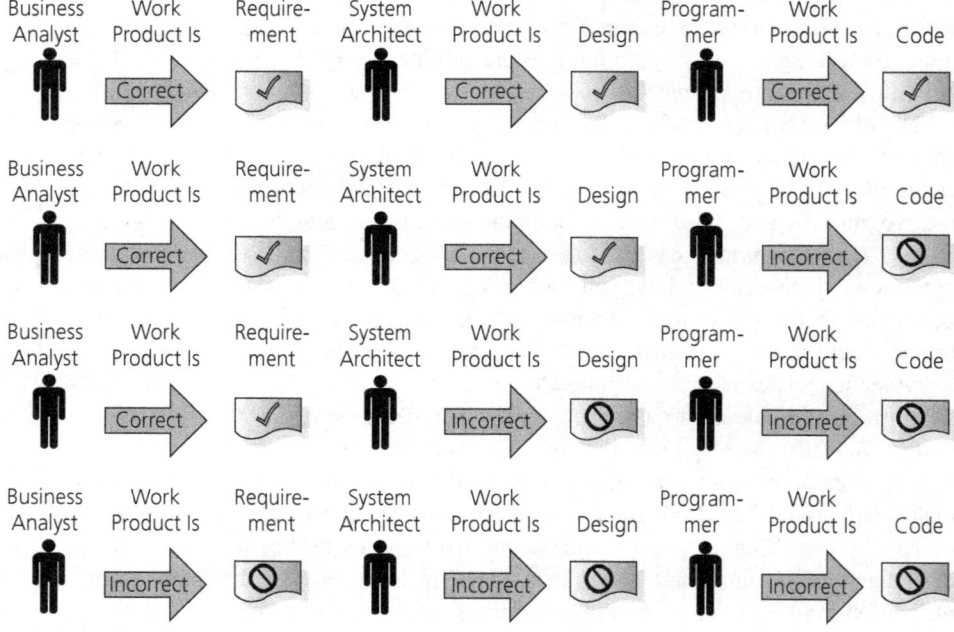

FIGURE 1.1   Four typical scenarios

Ideally, defects are removed in the same phase of the life cycle in which they are introduced. (Well, ideally defects are not introduced at all, but this is not possible because, as discussed before, people are fallible.) The extent to which defects are removed in the phase of introduction is called phase containment. Phase containment is important because the cost of finding and removing a defect increases each time that defect escapes to a later life cycle phase. Multiplicative increases in cost, of the sort seen in Figure 1.2, are not unusual. The specific increases vary considerably, with Boehm reporting cost increases of 1:5 (from requirements to after release) for simple systems, to as high as 1:100 for complex systems [Boehm 1986]. If you are curious about the economics of software testing and other quality-related activities, you can refer to Gilb [1993], Black [2004] or Black [2009]. Note, that this is also true in an Agile context. Teams that are spending more time on user story/acceptance test refinement sessions are much more efficient than others. Also finding problems at team level is much more efficient than at later stages (e.g. at user acceptance test level, which is sometimes organized outside of the Agile team).

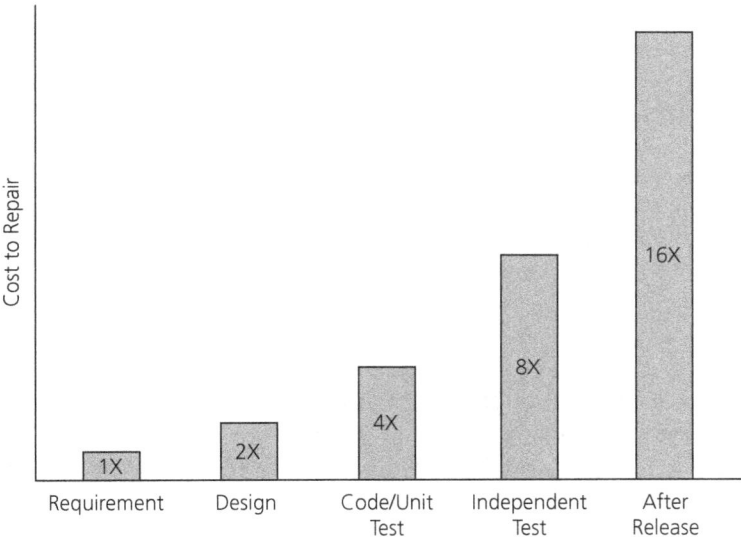

FIGURE 1.2 Multiplicative increases in cost

Defects may result in failures, or they may not, depending on inputs and other conditions. In some cases, a defect can exist that will never cause a failure in actual use, because the conditions that could cause the failure can never arise. In other cases, a defect can exist that will not cause a failure during testing, but which always results in failures in production. This can happen with security, reliability and performance defects, especially if the test environments do not closely replicate the production environment(s).

It can also happen that expected and actual results do not match for reasons other than a defect. In some cases, environmental conditions can lead to unexpected results that do not relate to a software defect. Radiation, magnetism, electronic fields and pollution can damage hardware or firmware, or simply change the conditions of the hardware or firmware temporarily in a way that causes the software to fail.

Defects, root causes and effects

Testing also provides a learning opportunity that allows for improved quality if lessons are learned from each project. If root cause analysis is carried out for the defects found on each project, the team can improve its software development processes to avoid the introduction of similar defects in future systems. Through

Root cause A source of a defect such that if it is removed, the occurrence of the defect type is decreased or removed.

this simple process of learning from past mistakes, organizations can continuously improve the quality of their processes and their software. A **root cause** is generally an organizational issue, whereas a cause for a defect is an individual action. So, for example, if a developer puts a 'less than' instead of 'greater than' symbol, this error may have been made through carelessness, but the carelessness may have been made worse because of intense time pressure to complete the module quickly. With more time for checking their work, or with better review processes, the defect would not have got through to the final product. It is human nature to blame individuals when in fact organizational pressure makes errors almost inevitable.

A good example of the difference between defects, root causes and effects is: suppose that incorrect interest payments result in customer complaints. There is just a single line of code that is incorrect. The code was written for a user story that was ambiguous, so the developer interpreted it in a way that they thought was sensible (but it was wrong). How did the user story come to be ambiguous? In this example, the product owner misunderstood how interest was to be calculated, so was unable to clearly specify what the interest calculation should have been. This misunderstanding could lead to a lot of similar defects, due to ambiguities in other user stories as well.

The failure here is the incorrect interest calculations for customers. The defect is the wrong calculation in the code. The root cause was the product owner's lack of knowledge about how interest should be calculated, and the effect was customer complaints.

The root cause can be addressed by providing additional training in interest rate calculations to the product owner, and possibly additional reviews of user stories by interest calculation experts. If this is done, then incorrect interest calculations due to ambiguous user stories should be a thing of the past.

Root cause analysis is covered in more detail in two other ISTQB qualifications: Expert Level Test Management, and Expert Level Improving the Test Process.

1.3 SEVEN TESTING PRINCIPLES

SYLLABUS LEARNING OBJECTIVES FOR 1.3 SEVEN TESTING PRINCIPLES (K2)

FL-1.3.1 Explain the seven testing principles (K2)

In this section, we will review seven fundamental principles of testing that have been observed over the last 40+ years. These principles, while not always understood or noticed, are in action on most if not all projects. Knowing how to spot these principles, and how to take advantage of them, will make you a better tester.

In addition to the descriptions of each principle that follow, you can refer to Table 1.1 for a quick reference of the principles and their text as written in the Syllabus.

TABLE 1.1   Testing principles

Principle 1	**Testing shows the presence, not the absence, of defects**	Testing can show that defects are present but cannot prove that there are no defects. Testing reduces the probability of undiscovered defects remaining in the software but, even if no defects are found, testing is not a proof of correctness.
Principle 2	**Exhaustive testing is impossible**	Testing everything (all combinations of inputs and preconditions) is not feasible except for trivial cases. Rather than attempting to test exhaustively, risk analysis, test techniques and priorities should be used to focus test efforts.
Principle 3	**Early testing saves time and money**	To find defects early, both static and dynamic test activities should be started as early as possible in the software development life cycle. Early testing is sometimes referred to as 'shift left'. Testing early in the software development life cycle helps reduce or eliminate costly changes (refer to Chapter 3, Section 3.1).
Principle 4	**Defects cluster together**	A small number of modules usually contain most of the defects discovered during pre-release testing, or they are responsible for most of the operational failures. Predicted defect clusters, and the actual observed defect clusters in test or operation, are an important input into a risk analysis used to focus the test effort (as mentioned in Principle 2).
Principle 5	**Tests wear out**	If the same tests are repeated over and over again, eventually these tests no longer find any new defects. To detect new defects, existing tests and test data are changed and new tests need to be written. However, in some cases, such as automated regression testing, repeating tests has a beneficial outcome, which is confidence that there are few regression defects.
Principle 6	**Testing is context dependent**	Testing is done differently in different contexts. For example, safety-critical software is tested differently from an e-commerce mobile app. As another example, testing in an Agile project is done differently from testing in a sequential life cycle project (refer to Chapter 2, Section 2.1).
Principle 7	**Absence-of-defects is a fallacy**	Some organizations expect that testers can run all possible tests and find all possible defects, but Principles 2 and 1, respectively, tell us that this is impossible. Further, it is a fallacy to expect that *just* finding and fixing a large number of defects will ensure the success of a system. For example, thoroughly testing all specified requirements and fixing all defects found could still produce a system that is difficult to use, that does not fulfil the users' needs and expectations or that is inferior compared to other competing systems.

Principle 1. Testing shows the presence, not the absence of defects

As mentioned in the previous section, a typical objective of many testing efforts is to find defects. Many testing organizations that the authors have worked with are quite effective at doing so. One of our exceptional clients consistently finds, on average, 99.5% of the defects in the software it tests. In addition, the defects left undiscovered are less important and unlikely to happen frequently in production. Sometimes, it turns out that this test team has indeed found 100% of the defects that would matter to customers, as no previously unreported defects are reported after release. Unfortunately, this level of effectiveness is not common.

However, no test team, test technique or test strategy can guarantee to achieve 100% defect-detection percentage (DDP)—or even 95%, which is considered excellent. Thus, it is important to understand that, while testing can show that defects are present, it cannot prove that there are no defects left undiscovered. Of course, as testing continues, we reduce the likelihood of defects that remain undiscovered, but eventually a form of Zeno's paradox takes hold: each additional test run may cut the risk of a remaining defect in half, but only an infinite number of tests can cut the risk down to zero.

That said, testers should not despair or let the perfect be the enemy of the good. While testing can never prove that the software works, it can reduce the remaining level of risk to product quality to an acceptable level, as mentioned before. In any endeavour worth doing, there is some risk. Software projects—and software testing—are endeavours worth doing.

Principle 2. Exhaustive testing is impossible

This principle is closely related to the previous principle. For any significantly-sized system (anything beyond the trivial software constructed in first-year software engineering courses), the number of possible test cases is either infinite or so close to infinite as to be practically innumerable.

Infinity is a tough concept for the human brain to comprehend or accept, so let's use an example. One of our clients mentioned that they had calculated the number of possible internal data value combinations in the Unix operating system as greater than the number of known molecules in the universe by four orders of magnitude. They further calculated that, even with their fastest automated tests, just to test all of these internal state combinations would require more time than the current age of the universe. Even that would not be a complete test of the operating system; it would only cover all the possible data value combinations.

So, we are confronted with a big, infinite cloud of possible tests; we must select a subset from it. One way to select tests is to wander aimlessly in the cloud of tests, selecting at random until we run out of time. While there is a place for automated random testing, by itself it is a poor strategy. We'll discuss testing strategies further in Chapter 5, but for the moment let's look at two.

One strategy for selecting tests is risk-based testing. In risk-based testing, we have a cross-functional team of project and product stakeholders perform a special type of risk analysis. In this analysis, stakeholders identify risks to the quality of the system, and assess the level of risk (often using likelihood and impact) associated with each risk item. We focus the test effort based on the level of risk, using the level of risk to determine the appropriate number of test cases for each risk item, and also to sequence the test cases.

Another strategy for selecting tests is requirements-based testing. In requirements-based testing, testers analyze the requirements specification (which would be user stories in Agile projects) to identify test conditions. These test conditions inherit the priority of the requirement or user story they derive from. We focus the test effort based on the priority to determine the appropriate number of test cases for each aspect, and also to sequence the test cases.

Principle 3. Early testing saves time and money

This principle tells us that we should start testing as early as possible in order to find as many defects as possible. In addition, since the cost of finding and removing a defect increases the longer that defect is in the system, early testing also means we are likely to minimize the cost of removing defects.

So, the first principle tells us that we cannot find all the bugs, but rather can only find some percentage of them. The second principle tells us that we cannot run every possible test. The third principle tells us to start testing early. What can we conclude when we put these three principles together?

Imagine that you have a system with 1,000 defects. Suppose we wait until the very end of the project and run one level of testing, system test. You find and fix 90% of the defects. That still leaves 100 defects, which presumably will escape to the customers or users.

Instead, suppose that you start testing early and continue throughout the life cycle. You perform requirements reviews, design reviews and code reviews. You perform unit testing, integration testing and system testing. Suppose that, during each test activity, you find and remove only 45% of the defects—half as effective as the previous system test level. Nevertheless, at the end of the process, fewer than 30 defects remain. Even though each test activity was only 45% effective at finding defects, the overall sequence of activities was 97% effective. Note that now we are doing both static testing (the reviews) and dynamic testing (the running of tests at the different test levels). This approach of starting test activities as early as possible is also called 'shift left' because the test activities are no longer all done on the right-hand side of a sequential life cycle diagram, but on the left-hand side at the beginning of development. Although unit test execution is of course on the right side of a sequential life cycle diagram, improving and spending more effort on unit testing early on is a very important part of the shift-left paradigm.

In addition, defects removed early cost less to remove. Further, since much of the cost in software engineering is associated with human effort, and since the size of a project team is relatively inflexible once that project is underway, reduced cost of defects also means reduced duration of the project. This situation is shown graphically in Figure 1.3.

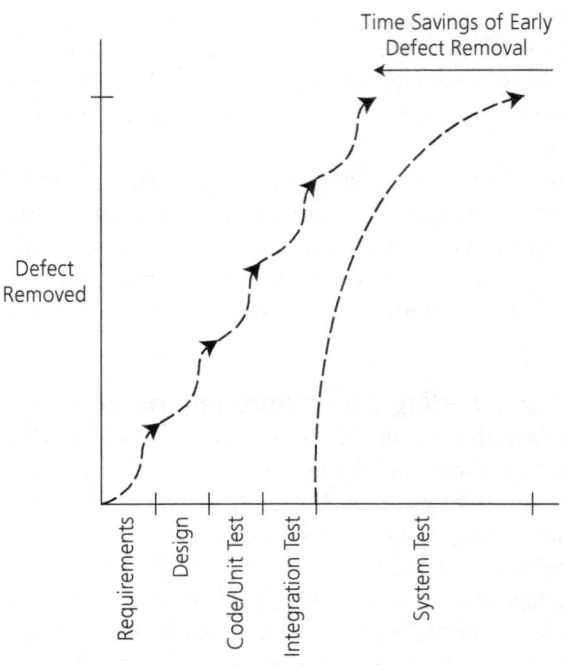

FIGURE 1.3   Time savings of early defect removal

Now, this type of cumulative and highly efficient defect removal only works if each of the test activities in the sequence is focused on different, defined objectives. If we simply test the same test conditions over and over, we will not achieve the cumulative effect, for reasons we will discuss in a moment.

Principle 4. *Defects cluster together*

This principle relates to something we discussed previously, that relying entirely on the testing strategy of a random walk in the infinite cloud of possible tests is relatively weak. Defects are not randomly and uniformly distributed throughout the software under test. Rather, defects tend to be found in clusters, with 20% (or fewer) of the modules accounting for 80% (or more) of the defects. In other words, the defect density of modules varies considerably. While controversy exists about why defect clustering happens, the reality of defect clustering is well established. It was first demonstrated in studies performed by IBM in the 1960s [Jones 2008], and is mentioned in Myers *et al.* [2011]. We continue to see evidence of defect clustering in our work with clients.

Defect clustering is helpful to us as testers, because it provides a useful guide. If we focus our test effort (at least in part) based on the expected (and ultimately observed) likelihood of finding a defect in a certain area, we can make our testing more effective and efficient, at least in terms of our objective of finding defects. Knowledge of and predictions about defect clusters are important inputs to the risk-based testing strategy discussed earlier. In a metaphorical way, we can imagine that bugs are social creatures who like to hang out together in the dark corners of the software.

Principle 5. *Tests wear out*

This principle was first noted by Boris Beizer [Beizer 1990], who coined the phrase 'pesticide paradox' to describe it. He observed that, just as a pesticide repeatedly sprayed on a field will kill fewer and fewer bugs each time it is used, so too will a given set of tests eventually stop finding new defects when re-run against a system under development or maintenance. The principle 'tests wear out' as such is thus based on the fact that repeating the same tests over and over again will result in finding fewer and fewer defects each time they are executed. If the tests do not provide adequate coverage, this slowdown in defect finding will result in a false level of confidence and excessive optimism among the project team. However, the air will be let out of the balloon once the system is released to customers and users.

Using the right test strategies is the first step towards achieving adequate coverage. However, no strategy is perfect. You should plan to regularly review the test results during the project, and revise the tests based on your findings. In some cases, you need to write new and different tests to exercise different parts of the software or system. These new tests can lead to discovery of previously unknown defect clusters, which is a good reason not to wait until the end of the test effort to review your test results and evaluate the adequacy of test coverage.

The principle that tests wear out is important when implementing the multilevel testing discussed previously in regard to the principle of early testing. Simply repeating our tests of the same conditions over and over will not result in good cumulative defect detection. However, when used properly, each type and level of testing has its own strengths and weaknesses in terms of defect detection, and collectively we can assemble a very effective sequence of defect filters from them. After such a sequence of complementary test activities, we can be confident that the coverage is adequate and that the remaining level of risk is acceptable. It should also be noted that of course tests don't 'wear out' like a physical object does and thereby stop finding defects that are there. Instead, because of the relatively low percentage of defects which are regression defects, there are simply fewer and fewer defects to find in the areas covered by the test.

It makes sense that the very first time any test is run, if there is a defect as yet undiscovered, that test will find it because it is the 'first on the scene'. The second time that same test is run, the defect may still be there, but it may have been fixed by now, so that test is less likely to find a defect. This is why exploratory testing is often very effective—all of the tests are new ones.

However, sometimes tests that are repeated over and over again can work in our favour, if it is not *new* defects that we are looking for. When we run automated regression tests, we are ensuring that the software that we are testing is still working as it was before, that is, there are no new unexpected side-effect defects that have appeared as a result of a change elsewhere. In this case, we are pleased that we have not found any new defects.

Principle 6. Testing is context dependent

Our safety-critical clients test with a great deal of rigour and care—and cost. When lives are at stake, we must be extremely careful to minimize the risk of undetected defects. Our clients who release software on the web, such as e-commerce sites, or who develop mobile apps can take advantage of the possibility to quickly change the software when necessary, leading to a different set of testing challenges—and opportunities. If you tried to apply safety-critical approaches to a mobile app, you might put the company out of business; if you tried to apply e-commerce approaches to safety-critical software, you could put lives in danger. So, the context of the testing influences how much testing we do and how the testing is done.

Another example is the way that testing is done in an Agile project as opposed to a sequential life cycle project. Every sprint in an Agile project includes testing of the functionality developed in that sprint; the testing is done by everyone on the Agile team (ideally) and the testing is done continually over the whole of development. In sequential life cycle projects, testing may be done more formally, documented in more detail and may be focused towards the end of the project.

Principle 7. Absence-of-defects fallacy

Throughout this section we have expounded the idea that a sequence of test activities, started early and targeting specific and diverse objectives and areas of the system, can effectively and efficiently find—and help a project team to remove—a large percentage of the defects. Surely that is all that is required to achieve project success?

Sadly, it is not. Many systems have been built that failed in user acceptance testing or in the marketplace, such as the initial launch of the US healthcare.gov website, which suffered from serious performance and web access problems.

Consider desktop computer operating systems. In the 1990s, as competition peaked for dominance of the PC operating system market, Unix and its variants had higher levels of quality than DOS and Windows. However, now in the 2020s, Windows dominates the desktop marketplace. One major reason is that Unix and its variants were too complex for most users in the early 1990s.

Consider a system that perfectly conforms to its requirements (if that were possible), which has been tested thoroughly and all defects found have been fixed. Surely this would be a success, right? Wrong! If the requirements were flawed, we now have a perfectly working wrong system. Perhaps it is hard to use, as in the previous example. Perhaps the requirements missed some major features that users were expecting or needed to have. Perhaps this system is quite OK, but a competitor has come out with a competing system that is easier to use, includes the expected features and is cheaper. Our 'perfect' system is not looking so good after all, even though it has effectively 'no defects' in terms of 'conformance to requirements'.

1.4 TEST ACTIVITIES, TESTWARE AND TEST ROLES

SYLLABUS LEARNING OBJECTIVES FOR 1.4 TEST ACTIVITIES, TESTWARE AND TEST ROLES (K2)

FL-1.4.1   **Summarize the different test activities and tasks (K2)**

FL-1.4.2   **Explain the impact of context on the test process (K2)**

FL-1.4.3   **Differentiate the testware that supports the test activities (K2)**

FL-1.4.4   **Explain the value of maintaining traceability (K2)**

FL-1.4.5   **Compare the different roles in testing**

In this section, we will describe the testing activities, tasks, work products and test roles. We will talk about the influence of context on the test process and the importance of traceability.

In this section, there are a large number of Glossary keywords: **coverage**, **test analysis**, **test basis**, **test case**, **test completion**, **test condition**, **test control**, **test data**, **test design**, **test execution**, **test implementation**, **test monitoring**, **test planning**, **test procedure**, **test result** and **testware**.

In Section 1.1, we looked at the definition of testing and identified misperceptions about testing, including that testing is not just test execution. Certainly, test execution is the most visible testing activity. However, effective and efficient testing requires test approaches that are properly planned and carried out, with tests designed and implemented to cover the proper areas of the system, executed in the right sequence and with their results reviewed regularly. This is a process, with tasks and activities that can be identified and need to be done, sometimes formally and other times very informally. In this section, we will look at the test activities and tasks in detail.

There is no 'one size fits all' test process, but testing does need to include common sets of activities, or it may not achieve its objectives. An organization may have a test strategy where the test activities are specified, including how they are implemented and when they occur within the life cycle. Another organization may have a test strategy where test activities are not formally specified, but expertise about test activities is shared among team members informally. The 'right' test process for you is one that achieves your test objectives in the most efficient way. The best test process for you would not be the best for another organization (and vice versa). Testing always needs to be tuned to its context.

Simply having a defined test strategy is not enough. A legal client of ours recently sued a company for a serious software failure. It turned out that while the company had a written test strategy, this strategy was not aligned with the testing best practices described in this book or the Syllabus. Further, upon close examination of their testware, it was clear that they had not even carried out the strategy properly or completely. The company ended up paying a substantial penalty for their lack of quality. So, you must consider whether your actual test activities and tasks are sufficient to meet your testing and quality objectives.

1.4.1 Test activities and tasks

A test process typically consists of the following main groups of activities:

- test planning
- test monitoring and control
- test analysis
- test design
- test implementation
- test execution
- test completion.

These activities appear to be logically sequential, in the sense that tasks within each activity often create the preconditions or precursor work products for tasks in subsequent activities. However, in many cases, the activities in the process may overlap or take place concurrently or iteratively, provided that these dependencies are fulfilled. Each group of activities consists of many individual tasks; these will vary for different projects or releases. For example, in Agile development, we have small iterations of software design, build and test that happen continuously, and planning is also a very dynamic activity throughout. If there are multiple teams, some teams may be doing test analysis while other teams are in the middle of test implementation, for example.

Note that this is 'a' test process, not 'the' test process. We have found that most of these activities, and many of the tasks within these activities, are carried out in some form or another on most successful test efforts. However, you should expect to have to tailor your test process, its main activities and the constituent tasks based on the organizational, project, process and product needs, constraints and other contextual realities. In sequential development, there will also be overlap, combination, concurrency or even omission of some tasks; this is why a test process is tailored for each project.

Test planning

Test planning The activity of establishing or updating a test plan.

Test planning involves defining the objectives of testing and the approach for meeting those objectives within project constraints and contexts. This includes deciding on suitable test techniques to use, deciding what tasks need to be done, formulating a test schedule and other things.

Metaphorically, you can think of test planning as similar to figuring out how to get from one place to another (without using your GPS—there is no GPS for testing). For small, simple and familiar projects, finding the route merely involves taking an existing map, highlighting the route and jotting down the specific directions. For large, complex or new projects, finding the route can involve a sophisticated process of creating a new map, exploring unknown territory and blazing a fresh trail.

We will discuss test planning in more detail in Section 5.1.

Test monitoring and control

Test monitoring The activity that checks the status of testing activities, identifies any variances from planned or expected, and reports status to stakeholders.

To continue our metaphor, even with the best map and the clearest directions, getting from one place to another involves careful attention, watching the dashboard, minor (and sometimes major) course corrections, talking with our companions about the journey, looking ahead for trouble, tracking progress towards the ultimate destination and coping with finding an alternate route if the road we wanted is blocked. So, in **test monitoring**, we continuously compare actual progress against the plan, check on the progress of test activities and report the test status and any necessary deviations from the plan. In **test control**, we take whatever actions are necessary to meet the mission and objectives of the project, and/or adjust the plan.

Test control The activity that develops and applies corrective actions to get a test project on track when it deviates from what was planned.

Test monitoring is the ongoing comparison of actual progress against the **test plan**, using any test monitoring metrics that we have defined in the test plan. Test progress against the plan is reported to stakeholders in test progress reports or stakeholder meetings. One option that is often overlooked is that if things are going very wrong, it may be time to stop the testing or even stop the project completely. In our driving analogy, once you find out that you are headed in completely the wrong direction, the best option is to stop and re-evaluate, not continue driving to the wrong place.

Test plan Documentation describing the test objectives to be achieved and the means and the schedule for achieving them, organized to coordinate testing activities.

(Note that we have included the definition of test plan here, even though it is not listed in the Syllabus as a term that you need to know for this chapter; otherwise the definition of test planning is not very informative.)

One way we can monitor test progress is by using exit criteria, also known as 'definition of done' in Agile development. For example, the exit criteria for test execution might include:

- Checking test results and logs against specified coverage criteria (we have not finished testing until we have tested what we planned to test).
- Assessing the level of component or system quality based on test results and logs (e.g. the number of defects found or ease of use).
- Assessing product risk and determining if more tests are needed to reduce the risk to an acceptable level.

We will discuss test planning, monitoring and control tasks in more detail in Chapter 5.

Test analysis

In **test analysis**, we analyze the test basis to identify testable features and define associated **test conditions**. Test analysis determines 'what to test', including measurable coverage criteria. We can say colloquially that the test basis is everything upon which we base our tests. The test basis can include requirements, user stories, design specifications, risk analysis reports, the system design and architecture, interface specifications and user expectations.

In test analysis, we transform the more general testing objectives defined in the test plan into tangible test conditions. The way in which these are specifically documented depends on the needs of the testers, the expectations of the project team, any applicable regulations and other considerations.

Test analysis The activity that identifies test conditions by analyzing the test basis.

Test condition A testable aspect of a component or system identified as a basis for testing.

Test analysis includes the following major activities and tasks:

- Analyze the test basis appropriate to the test level being considered. Examples of a test basis include:
  - Requirement specifications, for example business requirements, functional requirements, system requirements, user stories, epics, use cases or similar work products that specify desired functional and non-functional component or system behaviour. These specifications say what the component or system should do and are the source of tests to assess functionality as well as non-functional aspects such as performance or usability.
  - Design and implementation information, such as system or software architecture diagrams or documents, design specifications, call flows, modelling diagrams (for example, UML (Unified Modelling Language) or entity-relationship diagrams), interface specifications or similar work products that specify component or system structure. Structures for implemented systems or components can be a useful source of coverage criteria to ensure that sufficient testing has been done on those structures.
  - The implementation of the component or system itself, including code, database metadata and queries, and interfaces. Use all information about any aspect of the system to help identify what should be tested.
  - Risk analysis reports, which may consider functional, non-functional and structural aspects of the component or system. Testing should be more thorough in the areas of highest risk, so more test conditions should be identified in the highest-risk areas.

- Evaluate the test basis and test items to identify various types of defects that might occur (typically done by reviews), such as:
  - ambiguities
  - omissions
  - inconsistencies
  - inaccuracies
  - contradictions
  - superfluous statements.

- Identify features and sets of features to be tested.

- Identify and prioritize test conditions for each feature, based on analysis of the test basis, and considering functional, non-functional and structural characteristics, other business and technical factors, and levels of risks.

- Capture bi-directional traceability between each element of the test basis and the associated test conditions. This traceability should be bi-directional (we can trace in both forward and backward directions) so that we can check which test basis elements go with which test conditions (and vice versa) and determine the degree of coverage of the test basis by the test conditions. Refer to Section 1.4.4 for more on traceability. Traceability is also very important for maintenance testing, as we will discuss in Chapter 2, Section 2.3.

How are the test conditions actually identified from a test basis? The test techniques, which are described in Chapter 4, are used to identify test conditions. Black-box techniques identify functional and non-functional test conditions, white-box techniques identify structural test conditions and experience-based techniques can identify other important test conditions. Using techniques helps to reduce the likelihood of missing important conditions and helps to define more precise and accurate test conditions.

Sometimes the test conditions identified can be used as test objectives for a test charter. In exploratory testing, an experience-based technique (refer to Chapter 4, Section 4.4.2), test charters, are used as goals for the testing that will be carried out in an exploratory way, that is, test design, execution and learning in parallel. When these test objectives are traceable to the test basis, the coverage of those test conditions can be measured.

One of the most beneficial side effects of identifying what to test in test analysis is that you will find defects, for example inconsistencies in requirements, contradictory statements between different documents, missing requirements (such as no 'otherwise' for a selection of options) or descriptions that do not make sense. Rather than being a problem, this is a great opportunity to remove these defects before development goes any further. This verification (and validation) of specifications is particularly important if no other review processes for the test basis documents are in place.

Test analysis can also help to validate whether the requirements properly capture customer, user and other stakeholder needs. For example, techniques such as behaviour-driven development (BDD) and acceptance test-driven development (ATDD) both involve generating test conditions (and test cases) from user stories. BDD focuses on the behaviour of the system and ATDD focuses on the user view of the system, and both techniques involve defining acceptance criteria. Since these acceptance criteria are produced before coding, they also verify and validate the user stories and the acceptance criteria. More about this can be found in Sections 2.1.3 and 4.5.3, and in the ISTQB Foundation Level Agile Tester qualification.

Test design

Test analysis addresses 'what to test' and **test design** addresses the question 'how to test', that is, what specific inputs and data are needed in order to exercise the software for a particular test condition. In test design, test conditions are elaborated (at a high level) in **test cases**, sets of test cases and other testware. Test analysis identifies general 'things' to test, and test design makes these general things specific for the component or system that we are testing.

Test design includes the following major activities:

- Design and prioritize test cases and sets of test cases.
- Identify the necessary **test data** to support the test conditions and test cases as they are identified and designed.
- Design the test environment, including set-up, and identify any required infrastructure and tools.
- Capture bi-directional traceability between the test basis, test conditions and test cases (refer also to Section 1.4.4).

As with the identification of test conditions, test techniques are used to derive or elaborate test cases from the test conditions. These are described in Chapter 4, where test analysis and test design are discussed in more detail.

Just as in test analysis, test design can also identify defects—in the test basis and in the existing test conditions. Because test design is a deeper level of detail, some defects that were not obvious when looking at test basis at a high level, may become clear when deciding exactly what values to assign to test cases. For example, a test condition might be to check the boundary values of an input field, but when determining the exact values, we realize that a maximum value has not been specified in the test basis. Identifying defects at this point is a good thing because if they are fixed now, they will not cause problems later.

Which of these specific tasks applies to a particular project depends on various contextual issues relevant to the project, and these are discussed further in Section 1.4.2.

Test implementation

In **test implementation**, we specify **test procedures** (or test scripts). This involves combining the test cases in a particular order, as well as including any other information needed for test execution. Test implementation also involves setting up the test environment and anything else that needs to be done to prepare for test execution, such as creating testware. Test design asks 'how to test', and test implementation asks 'do we now have everything in place to run the tests?'

Test implementation includes the following major activities:

- Develop and prioritize the test procedures and, potentially, create automated test scripts.
- Create test suites from the test procedures and automated test scripts (if any).
- Arrange the test suites within a test execution schedule in a way that results in efficient test execution.
- Build the test environment (possibly including test harnesses, service virtualization, simulators and other infrastructure items) and verify that everything needed has been set up correctly.

Test design The activity that derives and specifies test cases from test conditions.

Test case A set of preconditions, inputs, actions (where applicable), expected results and postconditions, developed based on test conditions.

Test data Data needed for test execution.

Test implementation The activity that prepares the testware needed for test execution based on test analysis and design.

Test procedure A sequence of test cases in execution order and any associated actions that may be required to set up the initial preconditions and any wrap-up activities post execution.

- Prepare test data and ensure that it is properly loaded in the test environment (including inputs, data resident in databases and other data repositories, and system configuration data).
- Verify and update the bi-directional traceability between the test basis, test conditions, test cases, test procedures and test suites (refer also to Section 1.4.4).

Ideally, all of these tasks are completed before test execution begins, because otherwise precious, limited test execution time can be lost on these types of preparatory tasks. One of our clients reported losing as much as 25% of the test execution period to what they called 'environmental shakedown', which turned out to consist almost entirely of test implementation activities that could have been completed before the software was delivered.

Note that although we have discussed test design and test implementation as separate activities, in practice they are often combined and done together.

Not only are test design and implementation combined, but many test activities may be combined and carried out concurrently. For example, in exploratory testing (refer to Chapter 4, Section 4.4.2), test analysis, test design, test implementation and test execution are done in an interactive way throughout an exploratory test session.

Test execution

Test execution The activity that runs a test on a component or system producing actual results.

Testware Work products produced during the test process for use in planning, designing, executing, evaluating and reporting on testing.

Test result The consequence/outcome of the execution of a test.

In **test execution**, the test suites that have been assembled in test implementation are run, according to the test execution schedule.

Test execution includes the following major activities:

- Record the identities and versions of all of the test items (parts of the test object to be tested), test objects (system or component to be tested), test tools and other **testware**.
- Execute the tests either manually or by using an automated test execution tool, according to the planned sequence.
- Compare actual **test results** with expected results, observing where the actual and expected results differ. These differences may be the result of defects, but at this point we do not know, so we refer to them as anomalies.
- Analyze the anomalies in order to establish their likely causes. Failures may occur due to defects in the code or they may be false-positives. (A false-positive is where a defect is reported when there is no defect.) A failure may also be due to a test defect, such as defects in specified test data, in a test document or the test environment, or simply due to a mistake in the way the test was executed.
- Report defects based on the failures observed (refer to Chapter 5, Section 5.5). A failure due to a defect in the code means that we can write a defect report. Some organizations track test defects (i.e. defects in the tests themselves), while others do not.
- Log the outcome of test execution (e.g. pass, fail or blocked). This includes not only the anomalies observed, and the pass/fail status of the test cases, but also the identities and versions of the software under test, test tools and testware.
- As necessary, repeat test activities when actions are taken to resolve discrepancies. For example, we might need to re-run a test that previously failed in order to confirm a fix (confirmation testing). We might need to run an updated test. We might also need to run additional, previously executed tests to see whether defects

have been introduced in unchanged areas of the software or to see whether a fixed defect now makes another defect apparent (regression testing).

- Verify and update the bi-directional traceability between the test basis, test conditions, test cases, test procedures and test results.

As before, which of these specific tasks applies to a particular project depends on various contextual issues relevant to the project; these are discussed further in Section 1.4.2.

Test completion

Test completion activities collect data from completed test activities to consolidate experience, testware and any other relevant information. Test completion activities should occur at major project milestones. These can include when a software system is released, when a test project is completed (or cancelled), when an Agile project iteration is finished (e.g. as part of a retrospective meeting), when a test level has been completed or when a maintenance release has been completed. The specific milestones that involve test completion activities should be specified in the test plan.

Test completion includes the following major activities:

- Check whether all defect reports are closed, entering change requests or product backlog items for any defects that remain unresolved at the end of test execution.
- Create a test summary report to be communicated to stakeholders.
- Finalize and archive the test environment, the test data, the test infrastructure and other testware for later reuse.
- Hand over the testware to the maintenance teams, other project teams and/or other stakeholders who could benefit from its use.
- Analyze lessons learned from completed test activities to determine changes needed for future iterations, releases and projects (i.e. perform a retrospective). Refer also to Section 2.1.6.
- Use the information gathered to improve test process maturity, especially as an input to test planning for future projects.

The degree and extent to which test completion activities occur, and which specific test completion activities do occur, depends on various contextual issues relevant to the project, which are discussed further in Section 1.4.2.

> **Test completion** The activity that makes testware available for later use, leaves test environments in a satisfactory condition and communicates the results of testing to relevant stakeholders.

1.4.2 Test process in context

There is no one right test process that applies universally; each organization needs to adapt their test process depending on their context. Testing is not performed in isolation; it's an integral part of the software development process. Also, stakeholders play an essential role, as testing is performed to assess whether or not their needs have been fulfilled. The factors that influence the particular test process include the following (this list is not exhaustive):

- Stakeholders: their needs, expectations and requirements determine to a large extent how testing will be organized. Also factors such as the number of stakeholders, their availability and willingness to cooperate, need to be taken into account.

- Team members: their testing knowledge and skills, level of expertise and availability. For example, exploratory testing is only effective with experienced testers and test automation requires specific knowledge and skills.

- Business domain: the criticality of the test object (e.g. a simple text-based note-taking mobile app versus medical devices), identified risks (the lower the risks, the less formal the process needs to be, and vice versa) and market needs. In many so-called safety-critical domains, specific (legal) regulations and standards apply.

- Software development life cycle model and project methodologies being used. An Agile project developing mobile apps will have quite a different test process from an organization producing medical devices such as pacemakers.

- Technical factors such as the type of software, complexity, product architecture and technology used. For example, large complex projects may consider more test levels and test types, e.g. have several types of integration testing, with a test process reflecting that complexity.

- Project constraints, including:
  - budgets and resources
  - timescales
  - complexity
  - contractual requirements.

- Organizational policies and practices.
- The availability of tools to support the testing activities.

All these factors will impact on how testing will be performed. They will impact the test process, activities and tasks to be performed, test strategy and approach, test techniques to be used, degree of test automation, required level of coverage, level of detail of test documentation, reporting structure, etc.

Coverage The degree to which specified coverage items are exercised by a test suite, expressed as a percentage.

Test basis The body of knowledge used as the basis for test analysis and design.

As mentioned, one aspect to consider is **coverage**. Coverage is a partial measure of the thoroughness of testing. When examining the **test basis**, which is whatever the tests are being derived from (such as a requirement, user story, design or even code), a number of things can be identified as coverage items (we can tell whether or not we have tested them). For example, system level testing may want to ensure that every user story has been tested at least once; integration level testing may want to ensure that every communication path has been tested at least once; and, in component testing, developers may want to ensure that every code module, branch or statement has been tested at least once. When a test exercises the coverage item (user story, communication path or code element), then that item has been covered. Coverage is the percentage of coverage items that were exercised in a given test run.

It is useful to know what you want to cover with your testing right from the start; coverage can act as a Key Performance Indicator (KPI) and help to measure the achievement of test objectives (if they are related to coverage).

In addition to the coverage items relating to specifications or code, we may also have environmental aspects that we want to cover. For example, tests for mobile apps may need to be tested on a number of mobile devices and configurations—these are also coverage items or coverage criteria. We could also consider coverage items of user personas, stakeholders or user success criteria. Measuring and reporting the coverage of these aspects can also give confidence to stakeholders that failures in operation would be less likely.

There is more about coverage in Section 4.3. Test processes are described in more detail in ISO/IEC/IEEE 29119-2 [2021].

1.4.3 Testware

Testware is created as output work products from the test activities described earlier in Section 1.4.1. 'Work products' is the generic name given to any form of documentation, informal communication or artefact that is used in testing (or indeed in development). When testing is more formal, the majority of work products may be written documentation; when testing is less formal, the corresponding work product may just be a scrap of paper, or a note in someone's mobile phone. 'Work product' is a general term covering any type of information needed to do the work (testing or development).

Testware (test work products) is created as part of the test process, and there is significant variation in the types of work products created, in the ways they are organized and managed, and in the names used for them. The work products described in this section are in the ISTQB Glossary of terms. More information about work products can be found in ISO/IEC/IEEE 29119-3 [2021].

Test work products can be captured, stored and managed in configuration management tools, or possibly in test management tools or defect management tools.

Test planning work products

You will not be surprised to find that test planning work products include test plans. There may be different test plans for different test levels. The test plan typically includes information about the test basis, to which all the other work products will be related via traceability information (refer to Section 1.4.4). Test plans also include entry and exit criteria (also known as definition of ready and definition of done) for the testing within their scope—the exit criteria are used during test monitoring and control. Other test planning work products, also often part of a test plan, include the test schedule and risk register. The test schedule typically has a list of activities, tasks or events of the test process, identifying their intended start and finish dates and/or times, and dependencies. The risk register is a list of risks together with their likelihood and impact and information about the risk mitigation approach (refer to Section 5.2).

Beware of what people call a 'test plan'; we have seen this name applied to any kind of test document, including test case specifications and test execution schedules. A test plan is a planning document—it contains information about what is intended to happen in the future and is similar to a project plan. It does not contain detail of test conditions, test cases or other aspects of testing.

The test plan needs to be understandable to those who need to know the information contained in it. The two-page cryptic diagram that was called a 'test plan' at one organization would not be the right sort of work product for other organizations.

Test plans can cover a whole project or be specific to a test level or type of testing. Test plans are covered in more detail in Chapter 5, Section 5.1.

Test monitoring and control work products

The work products associated with test monitoring and control typically include different types of test reports and also documentation of control directives and risk information. Test progress reports are produced on an ongoing and/or a regular basis to keep stakeholders updated about progress, on a weekly or monthly basis, for example. Test summary reports are produced at test completion milestones as a way of summarizing the testing for a particular unit of work or test project. Any test report needs to include details relevant to its intended audience, the date of the report and the time period covered by the report. Test reports may include test execution results once those are available, in summary form (e.g. number of tests run, failed, passed or blocked and number of defects raised of different severity), and the status compared to the exit criteria or definition of done.

Test monitoring and control work products should address project management concerns such as budget and schedule, task completion, resource allocation, usage and effort. If action needs to be taken on the basis of information reported in a test report, those actions should be summarized in the next report to ensure that the desired effect of the action has been achieved.

These work products are further explained in Chapter 5, Sections 5.2 and 5.3.

Test analysis work products

Test analysis work products include mainly (prioritized) test conditions, as this is the output of the test analysis activity. Each test condition is ideally traceable to the test basis (and vice versa). Defect reports about defects found in the test basis as a result of test analysis can also be considered a work product from test analysis.

Here is an example: the specification for an ordering system describes a sales discount feature when customers put in large orders. The test conditions might be to test all the discount values for various order values. The discount values would be coverage items—we want to make sure we have tested each of them at least once. The work product would be the list of test conditions, that is, the discount values.

Test conditions are further discussed in Chapter 4.

Test design work products

The main work products resulting from test design are (prioritized) test cases and sets of test cases that exercise the test conditions identified in test analysis. In exploratory testing, a test charter may be a test design work product.

Sometimes the test cases at this stage are still rather vague and high-level, that is, without concrete values for inputs and expected results. For example, a test case for the sales discount might be to set up four existing customers, one who orders only a small amount so does not qualify for a discount, and the other three who order enough to qualify for a discount at each of the three discount levels respectively.

Having the test cases at a high level means that we can use the same test case across multiple test cycles with different specific or concrete data. For example, one application may have discounts of 2%, 5% and 10%, and another may have discounts of 10%, 20% and 25%. Our high-level test case adequately documents the scope of the test even though the details will be different in each application. The test case is traceable to and from the test condition that it is derived from.

We have seen that high-level test cases can have advantages, but there are also some aspects that you need to be aware of with high-level test cases. For example, it may be difficult to reproduce the test exactly, or different testers may use different test data, so the test case is not exactly repeatable. A high-level test case is not directly automatable; the tool needs exact instructions and specific data in order to execute the test. The skill and domain knowledge of the tester is also critical; a junior new-hire with no domain knowledge may struggle to know what they are supposed to be doing, unless they are well supported by more experienced testers. These are not insurmountable problems, but they do need to be considered.

Test design work products may also include test data requirements, the requirements and design of the test environment and the identification of infrastructure and tools. The extent and way in which these are documented may vary significantly from project to project or from one company to another.

When deriving test cases from the test conditions, we may also find defects or improvements that we could make to the test conditions, so the test conditions themselves may be further refined during test design. In our sales discount example, in test analysis we identified the three discounts as test conditions, but in test design, by looking at the test cases, we identified the 'no discount' test condition—a discount of 0%.

Test cases are discussed further in Chapter 4.

Test implementation work products

Work products for test implementation include:

- test procedures and the sequencing of those procedures
- automated test scripts
- test suites
- test data
- a test execution schedule
- test environment elements, e.g. stubs, drivers, simulators and service virtualization.

At this point, because we are preparing for test execution, we need to further refine any high-level test cases into low-level test cases that use concrete and specific data, both for test inputs and test data. In our loyalty discount example, we would now need to specify the details of our existing customers, decide on exactly how much each order will come to and calculate the final amount they would pay, including the discount. So, for example, Mrs Smith puts in an order for $50.01. Because her order is over $50, she gets a 10% discount, so she pays $45.01. We calculate the expected result for the test using a test oracle—the source of what the correct answer should be (in this case simple arithmetic). We would also need to set up Mrs Smith and other customers in the database as part of the preconditions of running the test, and this would be included in the test procedure.

In exploratory testing, we may be creating work products for test design and test implementation while doing test execution; traceability may be more difficult in this case.

Test implementation may also create work products that will be used by tools, for example test scripts for test execution tools, and sometimes work products are created *by* tools, such as a test execution schedule. Service virtualization may also create test implementation work products.

As in test design, we may further refine test conditions (and high-level test cases) during test implementation. For example, by deciding on the concrete values for our sales discount example, we realize that a test condition we omitted was to consider two different ways of clients paying between $45.01 and $50.00 (that is, with or without a discount). This may not be important to include in our tests, but it is an additional test condition.

Test execution work products

Work products for test execution include:

- test logs, a record of relevant details about the execution of tests
- defect reports (refer to Chapter 5, Section 5.5)
- documentation of the status of individual test cases or test procedures (e.g. ready to run, passed, failed, blocked, deliberately skipped, etc.).

In test execution, we want to know what happened when the tests were run. We want to know about any problems encountered that may have blocked some tests from running, for example if the network was down for a time. We want to know whether or not we were able to execute all of the tests that we planned to execute (which is not necessarily all of the tests that we have designed or implemented).

When test execution is finished (or is stopped), we should be able to report test results based on traceability, so that if a set of tests that were all related to the same requirement or user story failed, we will know about that. We also want to know which requirements have passed all of their planned tests, and any that have not yet been tested, that is, the tests are still waiting to be run. This is particularly important when test execution is stopped rather than completed.

Stakeholders are more likely to appreciate the implications of failing tests if they can be related to user stories or requirements, rather than just being told that a certain number of tests passed or failed. This is why traceability is important.

Test completion work products

The work products for test completion include the following:

- test completion reports (refer to Chapter 5, Section 5.3.2)
- action items for improvement of subsequent projects or iterations (e.g. following an Agile retrospective meeting)
- documented lessons learned
- change requests, e.g. as product backlog items
- finalized testware.

Test completion work products give closure to the whole of the test process and should provide ongoing ideas for increasing the effectiveness and efficiency of testing within the organization in the future.

1.4.4 Traceability between the test basis and test work products

We have mentioned bi-directional traceability for all test work products covered in Section 1.4.3. Bi-directional means that we can trace, for example, a given requirement through test conditions and test cases to the test execution results, but we can also trace the test execution results back to test cases, test cases back to test conditions, and test conditions back to requirements. No matter what the work products are called, this cross-linking gives many benefits to the test process. We saw how traceability can aid in the measurement and reporting of test coverage in order to report coverage and defects related to requirements, which is more meaningful and of more value to stakeholders. Traceability also enables us to see what risks have been addressed by the testing done so far, and enables us to more easily identify the remaining residual risks. By tracing back to the risks, we can communicate with stakeholders about the implications of releasing the system with the current outstanding risks.

Good traceability also supports the following:

- Analyzing the impact of changes, whether to requirements or to the component or system.
- Making testing auditable, and being able to measure coverage.
- Meeting its governance criteria (where applicable).
- Improving the coherence of test progress reports and test summary reports to stakeholders, as described above.
- Relating the technical aspects of testing to stakeholders in terms that they can understand.
- Providing information to assess product quality, process capability and project progress against business goals.

You may find that your test management or requirements management tool provides support for traceability of work products; if so, make use of that feature. Some organizations find that they have to build their own management systems in order to organize test work products in the way that they want and to ensure that they have bi-directional traceability. However the support is implemented, it is important to have automated support for traceability—it is not something that can be sustained without tool support.

1.4.5 Roles in testing

The location of testers or of a test team within a project organization can vary widely. Similarly, there is wide variation in the roles that people within testing play. Some of these roles occur frequently, some infrequently. Two roles that are found within many organizations are those of the test manager and the tester, though the same people may play both roles at various points during the project. The activities and tasks performed by these two roles will vary, depending on the organization, the project and product context, and the skills of the individuals. Let's take a look at the work typically done in these roles, starting with the test manager.

Test manager role

A test manager tends to be the person tasked with overall responsibility for the test process and successful leadership of the test activities. The test manager role may be performed by someone who holds this as their full-time position (a professional test manager), or it might be done by a QA manager, project manager or a development manager. Regarding the last two people on this list, warning bells about independence should be ringing in your head now, in addition to thoughts about how we can ensure that such non-testers gain the knowledge and outlook needed to manage testing. The test manager tasks may also be done by a senior tester. It is even possible for one person to take on the roles of testing and test management at the same time. In Agile software development, some of the test management tasks may be handled by the Agile team. Tasks that span multiple teams or the entire organization may be performed by a test manager outside of the development team. Whoever is playing the role, expect them to plan, monitor and control the testing work.

A test manager may be a single person trying to ensure that all testing is done as well as it can be, or a test manager may have one or more teams of people reporting to them (for example, in larger organizations). Sometimes the person at the top of the hierarchy would be called a test coordinator or test coach, and the individual teams may be led by a test leader or lead tester. In any case, the test manager's role is to manage the testing!

Typical tasks of the test manager role may include the following:

- At the outset of the project, in collaboration with the other stakeholders, devise the test objectives, organizational test policies (if not already in place) and test strategies.
- Plan the test activities, based on the test objectives and risks, and the context of the organization and the project. This may involve selecting the test approaches, estimating time, effort and cost for testing, acquiring resources, defining test levels, types and test cycles, and planning defect management.
- Write and update over time any test plan(s).
- Coordinate the test plan(s) with other project stakeholders, project managers, product owners and anyone else who may affect or be affected by the project or the testing.
- Share the testing perspective with other project activities, such as integration planning, especially where third-party suppliers are involved.
- Lead, guide and monitor the analysis, design, implementation and execution of the tests, monitor test progress and results, and check the status of exit criteria (or definition of done).
- Prepare and deliver test progress reports and test summary reports, based on information gathered from the testers.
- Adapt the test planning based on test results and progress (whether documented in test progress or summary reports or not) and take any actions necessary for test control.
- Support setting up the defect management system and adequate configuration management of the testware, and traceability of the tests to the test basis.
- Produce suitable metrics for measuring test progress and evaluating the quality of the testing and the product (test object).
- Recognize when test automation is appropriate and, if it is, plan and support the selection and implementation of tools to support the test process, including setting a budget for tool selection (and possible purchase, lease, support and training of the team), allocating time and effort for pilot projects and providing continuing support in the use of the tool(s). (Refer to Chapter 6 for more on tool support for testing.)
- Decide about the implementation of test environment(s) and ensure they are put into place before test execution and managed during test execution.
- Promote and advocate the testers, the test team and the test profession within the organization.
- Develop the skills and careers of testers, through training, performance evaluations, coaching and other activities, such as lunch-time discussions or presentations.

Although this is a list of typical activities of a test manager, the tasks may be carried out by people who are not labelled as a test manager. For example, in Agile development, the Agile team may perform some of these tasks, especially those to do with day-to-day testing within the team. Test managers who are outside an individual development team, who work with several teams or the whole organization, may be called a test coach. Refer to Black [2009] for more on managing the test process.

Tester role

As with the test manager role, projects should include testers at the outset. In the planning and preparation of the testing, testers should review and contribute to test plans, as well as analyzing, reviewing and assessing requirements and design specifications. They may be involved in or even be the primary people identifying test conditions and creating test designs, test cases, test procedure specifications and test data, and may automate or help to automate the tests. They often set up the test environments or assist system administration and network management staff in doing so.

In sequential life cycles, as test execution begins, the number of testers often increases, starting with the work required to implement tests in the test environment. They may play such a role on all test levels, even those not under the direct control of the test group, for example they might implement unit tests which were designed by developers. Testers execute and log the tests, evaluate the results and document problems found. They monitor the testing and the test environment, often using tools for this task, and often gather performance metrics. Throughout the testing life cycle, they review each other's work, including test specifications, defect reports and test results.

In Agile development, testers are involved in every iteration or sprint, performing a wide variety of testing tasks on whatever is being developed at the time, so they are continuously involved throughout. They may be reviewing user stories and identifying test conditions, implementing the tests, running them and reporting on results.

Testers may have specializations such as test analysis, test design, specific test types (especially non-functional testing) or test automation. Such specialists may take the role of tester at different test levels and at different times. For example, the person doing the testing tasks at component or component integration level is often the developer (but refer to caveats about independence in Section 1.5.3). At system testing or system integration testing, an independent test team may do the testing activities. In acceptance testing, the test tasks may be done by business analysts, subject matter experts, users or product owners. At operational acceptance testing, operations and/or system administration staff may be performing the tester tasks.

Typical tasks of the tester may include the following:

- Reviewing and contributing to test plans from the tester perspective.
- Analyzing, reviewing and assessing requirements, user stories and acceptance criteria, specifications and models (that is, the test basis) for testability and to detect defects early.
- Identifying and documenting test conditions and test cases, capturing traceability between test cases, test conditions and the test basis to assist in checking the thoroughness of testing (coverage), the impact of failed tests and the impact on the tests of changes in the test basis.

- Designing, setting up and verifying test environment(s), coordinating with system administration and network management.
- Designing and implementing test cases and test procedures, including automated tests where appropriate.
- Acquiring and preparing test data to be used in the tests.
- Creating a detailed test execution schedule (for manual tests).
- Executing the tests, evaluating the results and documenting deviations from expected results as defect reports.
- Using appropriate tools to help the test process.
- Automating tests as needed (for technical test specialists), as supported by a test automation engineer or expert or a developer. (Refer to Chapter 6 for more on test automation.)
- Evaluating non-functional characteristics such as performance efficiency, reliability, usability, security, compatibility and portability.
- Reviewing tests developed by others, including other testers, business analysts, developers or product owners. Part of a tester's role is to help educate others about doing better testing.

1.5 ESSENTIAL SKILLS AND GOOD PRACTICES IN TESTING

> **SYLLABUS LEARNING OBJECTIVES FOR 1.5 ESSENTIAL SKILLS AND GOOD PRACTICES IN TESTING (K2)**
>
> **FL-1.5.1** **Give examples of the generic skills required for testing (K2)**
>
> **FL-1.5.2** **Recall the advantages of the whole team approach (K1)**
>
> **FL-1.5.3** **Distinguish the benefits and drawbacks of independence of testing (K2)**

In this section, we will describe the generic skills required for testing and provide examples of various skills types. We will talk about why independence is important in testing and discuss the benefits and drawbacks associated with independent testing.

1.5.1 Generic skills required for testing

Doing testing properly requires more than defining the right positions and number of people for those positions. Good test teams have the right mix of skills based on the tasks and activities they need to carry out, and people outside the test team who are in charge of test tasks need the right skills, too.

People involved in testing need basic professional and social qualifications such as literacy, the ability to prepare and deliver written and verbal reports, the ability to communicate effectively and so on. Going beyond that, when we think of the skills that testers need, six main areas are particularly relevant for testers:

- **Application or business domain knowledge:** A tester must understand the intended behaviour, the problem the system will solve, the process it will automate and so forth, in order to spot improper behaviour while testing, and recognize the 'must-work' functions and features. Domain knowledge can make the testers identify the most interesting test cases from a business perspective. It will also make the tester better understand and be able to communicate effectively with end users and business representatives.

- **Technology knowledge:** A tester must be aware of issues, limitations and capabilities of the chosen implementation technology, in order to effectively and efficiently locate problems and recognize the 'likely-to-fail' functions and features. One area that technology relates to is test automation and test tooling. Being able to use the appropriate test tools allows the tester to increase the efficiency of testing.

- **Testing knowledge:** A tester must know the testing topics discussed in this book, and often more advanced testing topics, in order to effectively and efficiently carry out the test tasks assigned. Testing knowledge and skill will typically increase the effectiveness of testing, e.g. by using test techniques.

- **Personal characteristics:** Good testers tend to be curious, but also methodical, thorough and careful, paying close attention to detail, so that defects that are found are recorded in a way that developers can identify when they are fixed. Testers like to challenge things—and they should—especially if details are missing or ambiguous.

- **Analytical thinking:** The best 'tool' that a tester has is their brain. Critical thinking is essential in test analysis and design, and so is creativity, so that testing is as effective and efficient as possible (within project constraints).

- **Soft skills:** Any software tester should also possess so-called soft skills (also known as people skills). Soft skills are particularly important for testers due to the nature of what testers find in their work. Testers have an instinct and understanding for where and how software might fail, and how to find defects. A tester also should have the soft skills to influence and communicate in a manner in which they become vital to the project. Testing requires a toolbox full of soft skills including communication, time management, analytical, eager to learn and critical thinking but also relatively standard people skills such as reading, reporting and presentation skills. Some of these skills allow a tester to be better at finding defects, but most of them relate to being better at communicating often difficult messages. Testers are often the bearers of bad news. This makes communication skills crucial for testers. Communicating test results may be perceived as criticism of the product *and* of its author. Having these soft skills is like a prerequisite to having the right attitude for being an open-minded tester within an Agile team, for example having empathy towards other disciplines, knowledge sharing and being a team player. The specific skills in each area and the level of skill required vary by project, organization, application and the risks involved.

The set of testing tasks and activities are many and varied, and so too are the skills required, so we often see specialization of skills and separation of roles. For example, due to the special knowledge required in the areas of testing, technology and business domain, respectively, test automation experts may handle automating the regression tests, developers may perform component and integration tests, and users and operators may be involved in acceptance tests.

We have long advocated pervasive testing, the involvement of people throughout the project team in carrying out testing tasks. Let's close this section, though, on a cautionary note. Software and system companies (for example, producers of shrink-wrapped software and consumer products) typically overestimate the technology knowledge required to be an effective tester. Businesses that use information technology (for example, banks and insurance companies) typically overestimate the business domain knowledge needed.

All types of projects tend to underestimate the testing knowledge required. We have seen a project fail in part because people without proper testing skills were responsible for testing critical components, leading to the disastrous discovery of fundamental architectural problems later. Most projects can benefit from the participation of professional testers, as amateur testing alone will usually not suffice.

1.5.2 Whole team approach

The whole team approach is a fundamental concept in Agile methodologies, emphasizing collaboration and collective responsibility among all team members to deliver high-quality software and meet customer needs. An important skill for a tester is the ability to work effectively in a team context and to contribute to the team goals. The basic idea of the whole team approach in Agile development is that we are going to identify the necessary knowledge and skills required to complete the project, and then make sure that the team has those skills and knowledge. That includes the testers, who are a part of the team. The size of the team varies from as low as three to a maximum of about nine. For some of us engaged in consulting with Agile clients, we're finding that the five-to-nine range is pretty typical.

Ideally, the team should be co-located, to fulfil the principle of maximizing the information shared in face-to-face communications. People seem to have varying opinions about co-location. At a conference in Kyiv, one of the authors heard one of the Agile founders, Alistair Cockburn, say categorically that if you're not co-located, you're not doing Agile. However, some of us have worked with clients where the teams are not at all co-located, who say they're doing Agile, and it sure looks a lot like Agile. And it's working for them. So, they might not really care whether somebody from the outside says that what they're doing can't possibly be working for them.

Co-located or distributed, the team is supposed to have what are called daily stand up meetings. The daily stand up meetings are called stand up meetings because they're supposed to be short, about 15 minutes long, and everybody stands up. However, these meetings are another place where some amount of dysfunction within organizations can arise, where stand-up meetings are not stand-up meetings, they are sit down meetings and they last for hours. At one of our clients, we saw that these meetings would last for a couple of hours, and in these meetings they were redefining things constantly. That's not particularly Agile; that's chasing your tail.

In an ideal sequential life cycle, following V-model theory, everybody is responsible for quality, and quality is supposed to be built into the product from the start. That's the theory, anyway. The usual sequential life cycle practice has been that nobody's really looking at quality until the end of the project, when you get into system test and system integration test. Then, suddenly, there's an explosion of bugs. Quality is relegated to the tester's problem, but the late discovery of a large number of bugs turns out to be everyone's problem.

In Agile, you have an opportunity to break that dysfunction. One way you do that is by working closely with the business representatives to define the acceptance criteria and the acceptance tests as part of the user stories. Another way you do that is by working with the developers to make sure that what they're doing from a unit testing point of view is adequate. In this way, all project team members collaborate together to produce the highest quality product.

This collaboration is sometimes referred to as the power of three. Another phrase you'll hear is the three amigos. Amigo is the Spanish word for friend. So, the three amigos—the three friends—are the business stakeholder, the developer and the tester. Their different perspectives allow them to share their mutual knowledge. Communicating face to face allows them to make decisions without creating heavy weight documentation. Through this collaboration, they create a good set of tests that verify the requirements and they assist the user in validating that the code actually solves the problem. Developers and testers also collaborate on test automation.

Finally, a word about what the whole-team approach doesn't mean—or at least shouldn't mean. Some people have construed the whole team approach to mean that we don't need any independent testers. A number of our clients using Agile methods are struggling to figure out how to make independent testing work. There are two approaches that have not worked well with our clients, but there is a way to achieve good results.

The first unsuccessful model is not to have any independent test team at all, and simply to embed testers within each Agile team. Most of our clients have figured out that this doesn't work very well, for all the usual reasons. Remember, there's no magic in Agile methods, and they don't change human nature. What happens is that over time, the tester's independence and objectivity can be shaved away. I know some Agile testers will object to that statement, saying, 'Oh no, I'm still objective'. However, it's difficult when you're in a team, and you're supposed to collaborate, and there's supposed to be team work, to be the spoil-sport. It's difficult to be the one who comes in and says, 'Well, we've found some serious defects and we don't think this user story is ready to release'. A tester embedded within a development team will tend to adopt the prevailing ethos and objectives of that team, which may well be focused on velocity, especially if the business stakeholders have very powerful personalities.

To get the benefits of independence, some of our clients maintain completely independent and separated test teams, and one or two testers are assigned in the last few days or week of an iteration. They are like testing paratroopers, parachuting in to test the software. The objective is to ensure complete independence, impartiality and quality-focus. The trouble is that this approach delays the start of testing unnecessarily, creating time pressure and reducing test coverage. Further, the testers don't have the context, and often there's not enough documentation to ramp them up quickly. In addition, they don't have good relationships and are not seen as part of the team. Instead, they're seen as quality cops imposed on the team at the end. So, that model does not appear to work, at least for our clients who tried it.

What we've seen work is where a matrix approach is used, as shown in Figure 1.4. In this model, you have an independent test team, but testers are assigned to the teams on a long-term basis. This model is needed to deal with the risks that exist when using either of the other two options just mentioned. When testers are too close to the Agile team, they might lose the appropriate tester mindset. They might go along with the team and tolerate inefficient, ineffective or low-quality practices. However, when testers are outside the Agile team, they will find it hard to keep pace with change within time-box constraints.

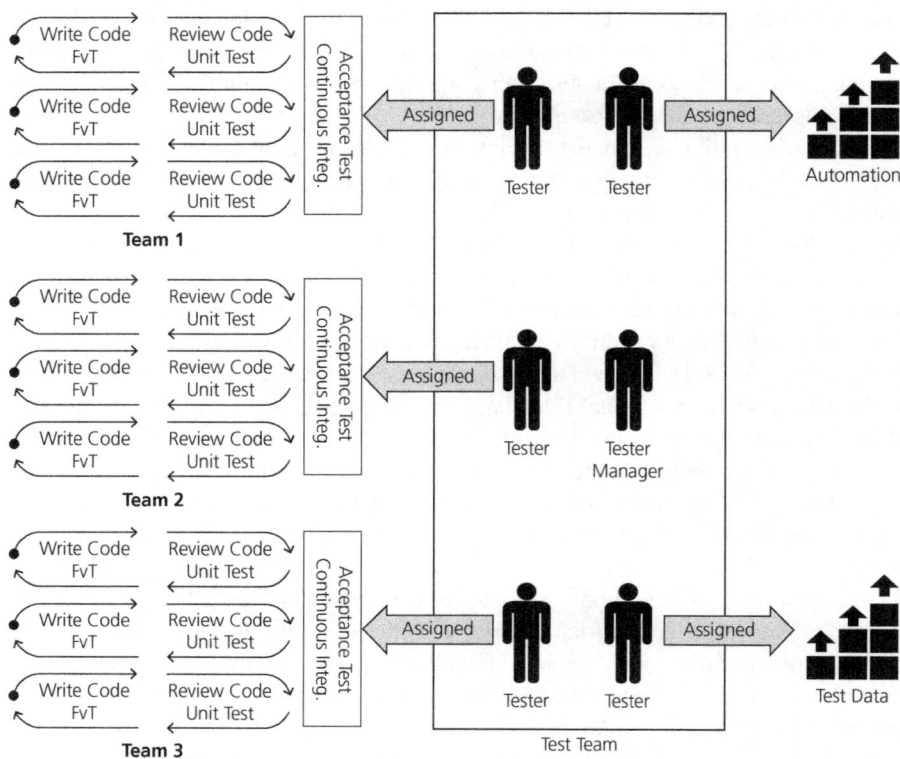

FIGURE 1.4   A matrix approach to testing

Figure 1.4 shows how some of our clients are successfully melding the whole-team approach and independent testing, getting the best of both worlds. The testers are assigned into the development teams, and they stay on the teams for a long period, at least six months, often longer. For the testers that are embedded within the teams, their day-to-day work is dictated by the needs of the teams. They work collaboratively with the whole team. However, if any team starts to lose focus on quality, to take shortcuts that endanger quality, or to make shortsighted release decisions, the tester has the right to escalate the problem to the test manager. A similar situation exists with the development managers. Both test manager and development manager act as coaches and supporters of the testers and developers in the development teams.

In addition to the testers in the teams, there are testers who are doing iteration-independent activities. These include building and maintaining test automation frameworks, running performance, security and system integration tests, creating and maintaining test data, creating and maintaining test environments, and so forth. Whether embedded in the development team or not, all the testers are part of the test team. This model is a matrix organizational structure, which is a well-proven management best practice. Note, that in some contexts, for example safety-critical, the whole team approach may not be the preferred option. In such situations a high level of test independence may be needed.

1.5.3 Independence of testing

Testing tasks may be done by people with a specific testing role, for example with tester as part of their job title, but testers are definitely not the only people who do testing. Developers, business analysts, users and customers also do testing tasks for different reasons and at different times. Even those who are full-time testers may do a variety of different tasks at different times. But if lots of people do testing, why have people dedicated to it? One reason is that the view of a different person, especially one who is trained to look for problems, can be much more effective at finding those problems. Many organizations, especially in safety-critical areas, have separate teams of independent testers to capitalize on this effect. As we discuss in the next (optional extra) section (1.6), independence can overcome cognitive bias.

If there is a separate test team, approaches to organizing it vary, as do the places in the organizational structure where the test team fits. Since testing is an assessment of quality, and since that assessment may not always be perceived as positive, many organizations strive to create an organizational climate where testers can deliver an independent, objective assessment of quality.

Levels of independence

When thinking about how independent the test team is, recognize that independence is not an either/or condition, but a continuum.

At one end of the continuum lies the absence of independence, where the developer performs testing on their own code within the development team. Of course, every good developer does do some testing of their own code, but this should not be the ONLY testing!

Moving towards independence, you find an integrated tester or group of testers working alongside the developers, but still within and reporting to the development manager. For example, developers may test each other's code after testing their own, or a tester on an Agile team may help developers and do some independent testing within the team. Pair programming is one way of having another pair of eyes on the code as it is being developed, whether it is two developers pairing, or a developer and a tester.

A further level of independence would be to have a team of testers who are independent and outside the development team, reporting to project management or business management.

Sometimes there are testers or teams of testers with special skills or responsibilities within an organization; such specialists may also be outside the development organization, which would be the other end of the independence continuum. For example, there may be a tester or team that specializes in performance or security testing.

In fully independent testing, you might see a separate test team reporting into the organization at a point equal to the development or project team or at a higher level. You might find specialists in the business domain (such as users of the system), specialists in technology (such as database experts) and specialists in testing (such as security testers or performance test experts) in a separate test team, as part of a larger independent test team, or as part of a contracted outsourced test team.

Potential benefits of independence

Let's examine the potential benefits and risks of independence, starting with the benefits.

An independent tester can often see more, different defects than a tester working within a development team—or a tester who is by profession a developer.

While business analysts, marketing staff, designers and developers bring their own assumptions to the specification and implementation of the item under test, an independent tester brings a different set of assumptions to testing and to reviews, which often helps expose hidden defects and problems related to the group's way of thinking, as we will discuss in more detail in Chapter 3. An independent tester brings a sceptical attitude of professional pessimism, a sense that, if there's any doubt about the observed behaviour, they should ask: 'Is this a defect?'

At the team level, an independent test team reporting to a senior or executive manager may enjoy (once they earn it) more credibility in the organization than a test manager or tester who is part of the development team. An independent tester who reports to senior management may be able to report their results honestly and without concern for reprisals that might result from pointing out problems in co-workers' or, worse yet, the manager's work. An independent test team often has a separate budget, which helps ensure the proper level of money is spent on tester training, testing tools, test equipment and so forth. In addition, in some organizations, testers in an independent test team may find it easier to have a career path that leads into more senior roles in testing.

Potential drawbacks of test independence

The use of independent test teams is not risk-free. It is possible for the testers and the test team to become isolated. This can take the form of interpersonal isolation from the developers, the designers and the project team itself. It can also take the form of isolation from the broader view of quality and the business objectives, for example an obsessive focus on defects, often accompanied by a refusal to accept business prioritization of defects. This leads to communication problems, feelings of alienation and antipathy, a lack of identification with and support for the project goals, spontaneous blame festivals and political backstabbing.

Even well-integrated test teams can suffer problems. Other project stakeholders might come to see the independent test team (rightly or wrongly) as a bottleneck and a source of delay.

Some developers abdicate their responsibility for quality, saying, 'Well, we have this test team now, so why do I need to unit test my code?'

Independent testers may not have all of the information they need about the test object, since they may be outside the development organization itself (where often much information is communicated informally). This leads to them being less effective than they should or could be.

Independence is not a replacement for familiarity. Although an independent view sees things that those closer to it miss, those who know the code or the test object best may be able to see some things that the independent tester would miss because of their limited knowledge. It is not a case of independence is always best, but of getting the best balance between independence and familiarity.

Independence varies

Due to a desire for the benefits of an independent test team, companies sometimes establish them, only to break them up again later. Why does that happen? A common cause is the failure of the test manager to effectively manage the risks of independence listed above. Some test teams succumb to the temptation to adopt a 'No can do' attitude, coming up with reasons why the project should bend to their needs rather than each side being flexible so as to enable project success. Testers take to acting as enforcers of process or as auditors, without a proper management mandate and support. Resentments and pressures build, until at last the organization decides that the independent test team causes more problems than it solves. It is especially important for testers and test managers to understand the mission they serve and the reasons why the organization wants an independent test team. Often, the entire test team must realize that, whether they are part of the project team or independent, they exist to provide a service to the project team.

There is no one right approach to organizing testing. For each project, you must consider whether to use an independent test team, based on the project, the application domain and the levels of risk, among other factors. As the size, complexity and criticality of the project increase, it is important to have independence in later levels of testing (like integration test, system test and acceptance test), though some testing is often best done by other people such as project managers, quality managers, developers, business and domain experts, or infrastructure or IT operations experts.

In most projects, there will be multiple test levels; the amount of independence in testing varies between levels as well (as it should). Often more independence is most effective at the higher levels, for example system and user acceptance testing.

The type of development life cycle also influences the level of independence of testing. In Agile development, a tester may provide some independence as part of the development team and may (also) be part of an independent team performing independent testing at higher levels. Product owners may perform acceptance testing to validate user stories at the end of each iteration.

1.6 THE PSYCHOLOGY OF TESTING (ADDITIONAL MATERIAL)

1.6.1 Human psychology and testing

In the previous Syllabus, there was a section about human psychology and testing. Although this has now been removed from the current v4.0 Syllabus, we think that this section, taken from the previous book, contains advice and information that may be useful to our readers, even though it is outside the Syllabus. In this section, we'll discuss the various psychological factors that influence testing and its success. We'll also contrast the mindset of a tester and a developer. For a good introduction to the psychology of testing, refer to Weinberg [2008].

As mentioned earlier, software testing includes both dynamic testing and static testing. Let's review the distinction again. Dynamic software testing involves actually executing the software or some part of it, such as checking an application-produced report for accuracy or checking response time to user input. Static software testing does not execute the software but uses two possible approaches: automated static analysis on the code (e.g. evaluating its complexity) or on a document (e.g. to evaluate the readability of a use case); or reviews of code or documents (e.g. to evaluate a requirement specification for consistency, ambiguity and completeness).

Finding defects
While dynamic and static testing are very different types of activities, they have in common their ability to find defects. Static testing finds defects directly, while dynamic testing finds evidence of a defect through a failure of the software to behave as expected. Either way, people carrying out static or dynamic tests must be focused on the possibility—indeed, the high likelihood in many cases—of finding defects. Indeed, finding defects is often a primary objective of static and dynamic testing activities.

Identifying defects may unfortunately be perceived by developers as a criticism, not only of the product but also of its author—and in a sense, it is. But finding defects in testing should be *constructive* criticism, where testers have the best interest of the developer in mind. One meaning of the word 'criticism' is 'an examination, interpretation, analysis or judgement about something'—this is an objective assessment. But other meanings include 'disapproval by pointing out faults or shortcomings' and even 'an attack on someone or something'. Testing does not want to be the latter sense of criticism, but even when intended in the first sense, it can be perceived in the other ways. Testers need to be diplomats, along with everything else.

Bias

However, there is another factor at work when we are reporting defects; the author (developer) believes that their code is correct—they obviously did not write it to be intentionally wrong. This confidence in their understanding is in some sense necessary for developers; they cannot proceed without it. But at the same time, this confidence creates confirmation bias. Confirmation bias makes it difficult to accept information that disagrees with your currently held beliefs. Simply put, the author of a work product has confidence that they have solved the requirements, design, metadata or code problem, at least in an acceptable fashion; however, strictly speaking, that is false confidence. Other biases may also be at work, and it is also human nature to blame the bearer of bad news (which defects are perceived to be).

Testers are not always aware of their biases either, and they do have biases of their own. Since those biases are different from the developers, that is a benefit, but a lack of awareness of those biases sets up potential conflict.

This reluctance to accept that their work is not perfect is why some people regard testing as a destructive activity (trying to destroy their work) rather than the constructive activity it is (trying to construct better quality software). Good testing contributes greatly to product quality and project quality, as we saw in Sections 1.1 and 1.2. To try to improve this view, information about defects and failures should be communicated in a constructive way.

While some developers are aware of their biases when they participate in reviews and perform unit testing of their own work products, those biases act to impede their effectiveness at finding their own defects. The mental mistakes that caused them to create the defects remain in their minds in most cases. When proofreading our own work, for example, we see what we meant, not what we wrote.

Review and test your own work?

Should software work product developers—business analysts, system designers, architects, database administrators and developers—review and test their own work? They certainly should; they have a deep understanding about the system, and quality is everyone's responsibility.

However, many business analysts, system designers, architects, database administrators and developers do not know the review, static analysis and dynamic testing techniques discussed in the Foundation Syllabus and this book. While that situation is gradually changing, much of the self-testing by software work product developers is either not done or is not done as effectively as it could be. The principles and techniques in the Foundation Syllabus and this book are intended to help either testers or others to be more effective at finding defects, both their own and those of others.

Attitudes

It is a particular problem when a tester revels in being the bearer of bad news. For example, one tester made a revealing—and not very flattering—remark during an interview with one of the authors. When asked what he liked about testing, he responded, 'I like to catch the developers'. He went on to explain that, when he found a defect in someone's work, he would go and demonstrate the failure on the programmer's workstation. He said that he made sure that he found at least one defect in everyone's work on a project, and went through this process of ritually humiliating the programmer with each and every one of their colleagues. When asked why, he said, 'I want to prove to everyone that I am their intellectual equal'. This person, while possessing many of the skills and traits you would want in a tester, had exactly the wrong personality to be a truly professional tester.

Instead of seeing themselves as their colleagues' adversaries or social inferiors out to prove their equality, testers should see themselves as teammates. In their special role, testers provide essential services in the development organization. They should ask themselves, 'Who are the stakeholders in the work that I do as a tester?' Having identified these stakeholders, they should ask each stakeholder group, 'What services do you want from testing, and how well are we providing them?'

While the specific services are not always defined, it is common that mature and wise developers know that studying their mistakes and the defects they have introduced is the key to learning how to get better. Furthermore, smart software development managers understand that finding and fixing defects during testing not only reduces the level of risk to the quality of the product, but it also saves time and money when compared to finding defects in production.

Communication

Clearly defined objectives and goals for testing, combined with constructive styles of communication on the part of test professionals, will help to avoid most negative personal or group dynamics between testers and their colleagues in the development team. Whenever defects are found, true testing professionals distinguish themselves by demonstrating good interpersonal skills. True testing professionals communicate facts about defects, progress and risks in an objective and constructive way that counteracts these misperceptions as much as possible. This helps to reduce tensions and build positive relationships with colleagues, supporting the view of testing as a constructive and helpful activity. While this is not necessary, we have noticed that many consummate testing professionals have business analysts, system designers, architects, developers and other specialists with whom they work as close personal friends.

This applies not only to testers but also to test managers, and not just to defects and failures but to all communication about testing, such as test results, test progress and risks.

Having good communication skills is a complex topic, well beyond the scope of a book on fundamental testing techniques. However, we can give you some basics for good communication with your development colleagues:

- Remember to think of your colleagues as teammates, not as opponents or adversaries. The way you regard people has a profound effect on the way you treat them. You do not have to think in terms of kinship or achieving world peace, but you should keep in mind that everyone on the development team has the common goal of delivering a quality system, and everyone must work together to accomplish that. Start with collaboration, not battles.

- Make sure that you focus on and emphasize the value and benefits of testing. Remind your developer colleagues that defect information provided by testing can help them to improve their own skills and future work products. Remind managers that defects found early by testing and fixed as soon as possible will save time and money and reduce overall product quality risk. Also, be sure to respond well when developers find problems in your own test work products. Ask them to review them and thank them for their findings (just as you would like to be thanked for finding problems in their work).

- Recognize that your colleagues have pride in their work, just as you do, and as such you owe them a tactful communication about defects you have found. It is not really any harder to communicate your findings, especially the potentially embarrassing findings, in a neutral, fact-focused way. In fact, you will find that if you avoid criticizing people and their work products, but instead keep your written and verbal communications objective and factual, you will also avoid a lot of unnecessary conflict and drama with your colleagues.

- Before you communicate these potentially embarrassing findings, mentally put yourself in the position of the person who created the work product. How are they going to feel about this information? How might they react? What can you do to help them get the essential message that they need to receive without provoking a negative emotional reaction from them?

- Keep in mind the psychological element of cognitive dissonance. Cognitive dissonance is a defect—or perhaps a feature—in the human brain that makes it difficult to process unexpected information, especially bad news. So, while you might have been clear in what you said or wrote, the person on the receiving end might not have clearly understood. Cognitive dissonance is a two-way street, too, and it is quite possible that you are misunderstanding someone's reaction to your findings. So, before assuming the worst about someone and their motivations, confirm that the other person has understood what you have said and vice versa.

1.6.2 Testers' and developers' mindsets

Testers and developers actually have different thought processes and different objectives for their work. A mindset reflects an individual's assumptions and the way that they like to solve problems and make decisions.

A tester's mindset should include the following:

- Curiosity: good testers are curious about why systems behave the way they do and how systems are built. When they see unexpected behaviour, they have a natural urge to explore further, to isolate the failure, to look for more generalized problems and to gain deeper understanding.

- Professional pessimism: good testers expect to find defects and failures. They understand human fallibility and its implications for software development. (However, this is not to say that they are negative or adversarial, as we'll discuss in a moment.)

- A critical eye: good testers couple professional pessimism with a natural inclination to doubt the correctness of software work products and their behaviours as they look at them. A good tester has, as a personal slogan, 'If in doubt, it is a bug'.

- Attention to detail: good testers notice everything, even the smallest details. Sometimes these details are cosmetic problems like font-size mismatches, but sometimes these details are subtle clues that a serious failure is about to happen. This trait is both a blessing and a curse. Some testers find that they cannot turn this trait off, so they are constantly finding defects in the real world—even when not being paid to find them.

- Experience: good testers not only know a defect when they see one, they also know where to look for defects. Experienced testers have seen a veritable parade of bugs in their time, and they leverage this experience during all types of testing, especially experience-based testing such as error guessing (refer to Chapter 4).

- Good communication skills: all of these traits are essential, but without the ability to effectively communicate their findings, testers will produce useful information that will, alas, be put to no use. Good communicators know how to explain the test results, even negative results such as serious defects and quality risks, without coming across as preachy, scolding or defeatist.

A good tester has the skills, the training, the certification and the mindset of a professional tester, and of these four skills, the most important—and perhaps the most elusive—is the mindset.

The tester's mindset is to think about what could go wrong and what is missing. The tester looks at a statement in a requirement or user story and asks, 'What if it isn't? What haven't they thought of here? What could go wrong?' That mindset is quite different from the mindset that a business analyst, system designer, architect, database administrator or developer must bring to creating the work products involved in developing software. While the testers (or reviewers) must assume that the work product under review or test is defective in some way—and it is their job to find those defects—the people developing that work product must have confidence that they understand how to do so properly. Looking at a statement in a requirement or user story, the developer thinks, 'How can I implement this? What technical challenges do I need to solve?'

Having a tester or group of testers who are organizationally separate from development, either as individuals or as an independent test team, can provide significant benefits, such as increased defect-detection percentage. A tester's mindset is a 'different pair of eyes' and independent testers can see things that developers do not see (because of confirmation bias discussed earlier). This is especially important for large, complex or safety-critical systems.

However, independence from the developers does not mean an adversarial relationship with them. In fact, such a relationship is toxic, often fatally so, to a test team's effectiveness.

The softer side of software testing is often the harder side to master. A tester may have adequate or even excellent technique skills and certifications, but if they do not have adequate interpersonal and communication skills, they will not be an effective tester. Such soft skills can be improved with training and practice. The best testers continuously strive to attain a more professional mindset, and it is a lifelong journey.

CHAPTER REVIEW

Let's review what you have learned in this chapter.

From Section 1.1, you should now know what testing is. You should be able to remember the typical objectives of testing. You should know the difference between testing and debugging. You should know the Glossary keyword terms **debugging**, **test object**, **test objective**, **testing**, **validation** and **verification**.

From Section 1.2, you should now be able to explain why testing is necessary and support that explanation with examples. You should be able to describe how testing contributes to the success of software development. You should be able to explain the difference between testing and QA and how they work together to improve quality. You should be able to distinguish between an error (made by a person), a defect (in a work product) and a failure (where the component or system does not perform as expected). You should know the difference between the root cause of a defect and the effects of a defect or failure. You should know the Glossary terms **defect**, **error**, **failure**, **quality**, **QA** and **root cause**.

You should be able to explain the seven principles of testing, discussed in Section 1.3.

From Section 1.4, you should now be able to recall the main testing activities of test planning, test monitoring and control, test analysis, test design, test implementation, test execution and test completion. You should know how context influences the test process. You should be familiar with the work products produced by each test activity, and the reasons and importance of good traceability. You should be able to compare the two principal roles in testing: the test management role and the testing role. You should know the Glossary terms **coverage**, **test analysis**, **test basis**, **test case**, **test completion**, **test condition**, **test control**, **test data**, **test design**, **test execution**, **test implementation**, **test monitoring**, **test planning**, **test procedure**, **test result** and **testware**.

From Section 1.5, you now should be able to give examples of the generic skills required for testing. You should be able to understand the whole-team approach and recall its advantages whereby an important skill for a tester is the ability to work effectively in a team context and to contribute to the team goals. You should know why independent testing is important, but also be able to analyze the potential benefits and problems associated with independence. There are no specific Glossary terms to know associated with this section.

SAMPLE EXAM QUESTIONS

Question 1 What is NOT a reason for testing?

a. To enable developers to code as quickly as possible.

b. To reduce the risk of failures in operation.

c. To contribute to the quality of the components or systems.

d. To meet any applicable contractual or legal requirements.

Question 2 Which one of the following best describes the difference between testing and debugging?

a. Testing shows failures that are caused by defects. Debugging finds, analyzes and removes the causes of failures in the software.

b. Testing finds defects. Debugging analyzes the defects and proposes preventive activities.

c. Testing removes defects. Debugging identifies the causes of failures.

d. Dynamic testing prevents causes of failures. Debugging removes the failures.

Question 3 Which statement about quality assurance (QA) is true?

a. QA and testing are the same.

b. QA includes both testing and root cause analysis.

c. Testing is quality control, not QA.

d. QA does not apply to testing.

Question 4 It is important to ensure that test design starts during the requirements definition. Which of the following test objectives supports this?

a. Preventing defects in the system.

b. Finding defects through dynamic testing.

c. Gaining confidence in the system.

d. Finishing the project on time.

Question 5 A test team consistently finds a large number of defects during development, including system testing. Although the test manager understands that this is good defect finding within their budget for their test team and industry, senior management and executives remain disappointed in the test group, saying that the test team missed some bugs that the users found after release. Given that the users are generally happy with the system and that the failures that have occurred have generally been low-impact, which of the following testing principles is most likely to help the test manager explain to these managers and executives why some defects are likely to be missed?

a. Exhaustive testing is impossible.

b. Defects cluster together.

c. Tests wear out.

d. Absence-of-defects fallacy.

Question 6 What are the benefits of traceability between the test basis and test work products?

a. Traceability means that test basis documents and test work products do not need to be reviewed.

b. Traceability ensures that test work products are limited in number to save time in producing them.

c. Traceability enables test progress and defects to be reported with reference to requirements, which is more understandable to stakeholders.

d. Traceability enables developers to produce code that is easier to test.

Question 7 Which of the following statements about the whole-team approach is true?

a. In the whole-team approach any team member can perform any task, but the tester is responsible for quality.

b. The whole-team approach helps ensure developers do not get too far ahead of the rest of the team, causing testing to lag behind.

c. The whole-team approach tends to advocate the idea of team members being 'generalists' without having deep skills in specific disciplines.

d. The whole-team approach is a consensus-based approach for estimating, mostly used to estimate the relative size of user stories.

Question 8 Given the following test work products, identify the testing activity that produces it.

1. Test schedule.
2. Test cases.
3. Test progress reports.
4. Defect reports.

a. 1) – Test planning, 2) – Test design 3) – Test execution, 4) – Test implementation.

b. 1) – Test execution, 2) – Test analysis 3) – Test completion, 4) – Test execution.

c. 1) – Test control, 2) – Test analysis, 3) – Test monitoring, 4) – Test implementation.

d. 1) – Test implementation, 2) – Test design, 3) – Test monitoring, 4) – Test execution.

Question 9 What is a potential drawback of independent testing?

a. Independent testing is likely to find the same defects as developers.

b. Developers may lose a sense of responsibility for quality.

c. Independent testing may find more defects.

d. Independent testers may be too familiar with the system.

Question 10 Which of the following is among the typical tasks of a test manager?

a. Develop system requirements, design specifications and usage models.

b. Handle all test automation duties.

c. Keep tests and test coverage hidden from developers.

d. Gather and report test progress metrics.

Question 11 According to the ISTQB Glossary, what do we mean when we call someone a test manager?

a. A test manager manages a team of junior testers.

b. A test manager plans and controls testing activities.

c. A test manager sets up test environments.

d. A test manager creates a detailed test execution schedule.

Question 12 What is the primary difference between the test plan, the test design and the test procedure?

a. The test plan describes one or more levels of testing, the test design identifies the associated high-level test cases and a test procedure describes the actions for executing a test.

b. The test plan is for managers, the test design is for developers and the test procedure is for testers who are automating tests.

c. The test plan is the least thorough, the test procedure is the most thorough and the test design specification is midway between the two.

d. The test plan is finished in the first third of the project, the test design is finished in the middle third of the project and the test procedure is finished in the last third of the project.

CHAPTER TWO

Testing throughout the software development life cycle

Testing is not a stand-alone activity. It has its place within a software development life cycle model and therefore the life cycle applied will largely determine how testing is organized. There are many different forms of testing. Because several disciplines, often with different interests, are involved in the development life cycle, it is important to clearly understand and define the various test levels and types. This chapter discusses the most commonly applied software development models, test levels and test types. Maintenance can be seen as a specific instance of a development process. The way maintenance influences the test process, levels and types and how testing can be organized is described in the last section of this chapter.

2.1 SOFTWARE DEVELOPMENT LIFE CYCLE MODELS

> **SYLLABUS LEARNING OBJECTIVES FOR 2.1 SOFTWARE DEVELOPMENT LIFE CYCLE MODELS (K2)**
>
> **FL-2.1.1** Explain the impact of the chosen software development lifecycle on testing (K2)
>
> **FL-2.1.2** Recall good testing practices that apply to all software development lifecycles (K1)
>
> **FL-2.1.3** Recall the examples of test-first approaches to development (K1)
>
> **FL-2.1.4** Summarize how DevOps might have an impact on testing (K2)
>
> **FL-2.1.5** Explain the shift-left approach (K2)
>
> **FL-2.1.6** Explain how retrospectives can be used as a mechanism for process improvement (K2)

In this section, we will discuss software development models and how testing fits into them. We will discuss sequential models, focusing on the V-model approach rather than the waterfall. We will discuss iterative development models (e.g. Spiral model, prototyping) and incremental models such as Unified Process. We will also look into Agile software development processes, methods and practices. Examples that will be discussed include Acceptance Test-Driven Development (ATTD), Behaviour-Driven Development (BDD), Extreme Programming (XP), Kanban, Scrum and Test-Driven Development (TDD).

As we go through this section, watch for the Syllabus terms **shift-left** and **test level**. You will find these keywords defined in the Glossary (and International Software Testing Qualifications Board (ISTQB) website).

The development process adopted for a project will depend on the project aims and goals. There are numerous development life cycles that have been developed in order to achieve different required objectives. These life cycles range from lightweight and fast methodologies, where time to market is of the essence, through to fully controlled and documented methodologies. Each of these methodologies has its place in modern software development, and the most appropriate development process should be applied to each project. The models specify the various stages of the process and the order in which they are carried out.

We will first look at the two main categories of software development life cycle models: sequential and iterative/incremental. Thereafter the Agile lifecycle model will be discussed together with some specific examples (Scrum and Kanban).

Sequential development models

Sequential development model A type of software development lifecycle model in which a complete system is developed in a linear way of several discrete and successive phases with no overlap between them.

A **sequential development model** is one where the development activities happen in a prescribed sequence—at least that is the idea. The models assume a linear sequential flow of activities; the next phase is only supposed to start when the previous phase is complete. Berard [1993] said that developing software from requirements is like walking on water—it is easier if it is frozen. But practice does not conform to theory: the activities will overlap, and things will be discovered in a later phase that may invalidate assumptions made in previous phases (which were supposed to be finished).

The waterfall model (in Figure 2.1) was one of the earliest models to be designed. It has a natural timeline where tasks are executed in a sequential fashion. We start at the top of the waterfall with a feasibility study and flow down through the various project tasks, finishing with implementation into the live environment. Design flows through into development, which in turn flows into build, and finally on into test. Different models have different levels; the figure shows one possible model. With all waterfall models, however, testing tends to happen towards the end of the life cycle, so defects are detected close to the live implementation date. With this model, it is difficult to get feedback passed backwards up the waterfall and there are difficulties if we need to carry out numerous iterations for a particular phase.

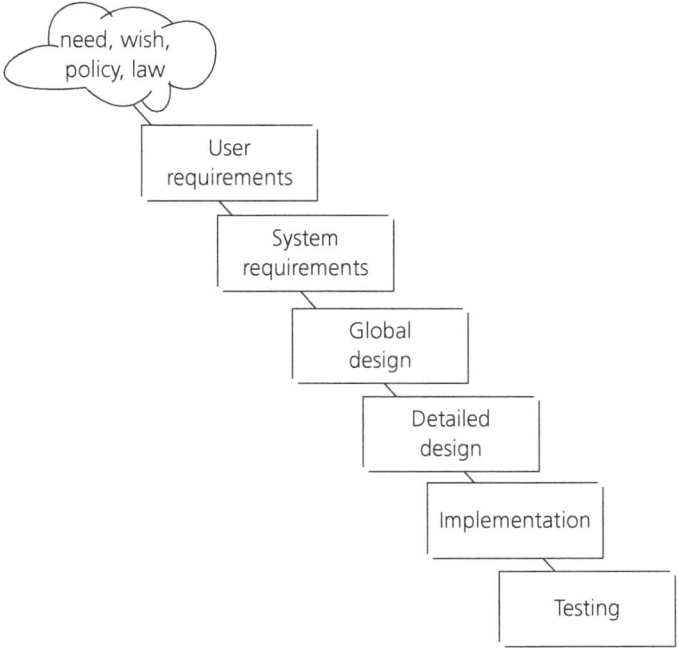

FIGURE 2.1   Waterfall model

The V-model was developed to address some of the problems experienced using the traditional waterfall approach. Defects were being found too late in the life cycle, as testing was not involved until the end of the project. Testing also added lead time due to its late involvement. The V-model provides guidance on how testing begins as early as possible in the life cycle. It also shows that testing is not only an execution-based activity. There are a variety of activities that need to be performed before the end of the coding phase. These activities should be carried out in *parallel* with development activities, and testers need to work with developers and business analysts so they can perform these activities and tasks, producing a set of test deliverables. The work products produced by the developers and business analysts during development are the basis of testing in one or more levels. By starting test design early, defects are often found in the test basis documents. A good practice is to have testers involved even earlier, during the review of the (draft) test basis documents. The V-model is a model that illustrates how testing activities (verification and validation) can be integrated into each phase of the life cycle. Within the V-model, validation testing takes place especially during the early stages, for example reviewing the user requirements, and late in the life cycle, for example during user acceptance testing.

Although variants of the V-model exist, a common type of V-model uses four **test levels**. (Refer to Section 2.2 for more on test levels.) The four test levels used, each with their own objectives, are:

Test level (test stage)
A specific instantiation
of a test process.

- Component testing: searches for defects in and verifies the functioning of software components (for example, modules, programs, objects, classes, etc.) that are separately testable.

- Integration testing: tests interfaces between components, interactions to different parts of a system such as an operating system, file system and hardware or interfaces between systems.

- System testing: concerned with the behaviour of the whole system/product as defined by the scope of a development project or product. The main focus of system testing is verification against specified requirements.

- Acceptance testing: validation testing with respect to user needs, requirements and business processes conducted to determine whether or not to accept the system.

In practice, a V-model may have more, fewer or different levels of development and testing, depending on the project and the software product. For example, there may be component integration testing after component testing and system integration testing after system testing. Test levels can be combined or reorganized depending on the nature of the project or the system architecture. In the V-model, there may also be overlapping of activities.

Sequential models aim to deliver all of the software at once, that is, the complete set of features required by stakeholders or users, or the software may be delivered in releases containing significant chunks of new functionality. However, typically this may take months or even years of development even for a single release.

Note that the types of work products mentioned in Figure 2.2 on the left side of the V-model are just an illustration. In practice, they come under many different names.

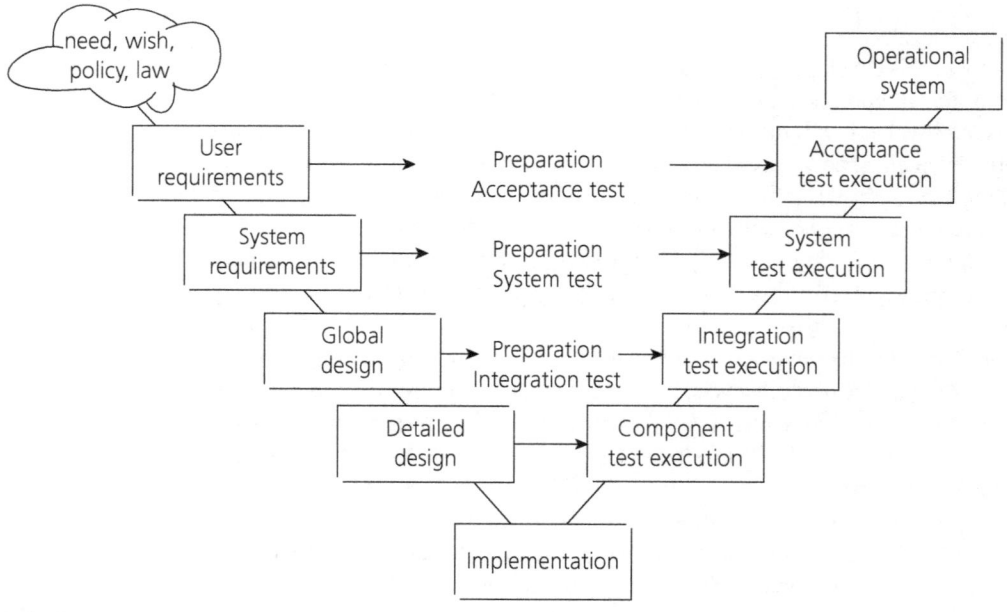

FIGURE 2.2 V-model

Iterative and incremental development models

Not all life cycles are sequential. There are also iterative and incremental life cycles where, instead of one large development timeline from beginning to end, we cycle through a number of smaller self-contained life cycle phases for the same project. As with the V-model, there are many variants of iterative and incremental life cycles.

To better understand the meaning of these two terms, consider the following two sequences of producing a painting.

- **Incremental:** complete one piece at a time (scheduling or staging strategy). Each increment may be delivered to the customer.

Images provided by: Mark Fewster, Grove Software Testing Ltd

- **Iterative:** start with a rough product and refine it, iteratively (rework strategy). Final version only delivered to the customer (although in practice, intermediate versions may be delivered to selected customers to get feedback).

Images provided by: Mark Fewster, Grove Software Testing Ltd

In terms of developing software, a purely iterative model does not produce a working system until the final iteration. The incremental approach produces working versions of parts of the system early on and each of these can be released to the customer. The advantage of this is that the customer can gain early benefit from using the deliveries and, perhaps most importantly, the customers can give valuable feedback. This feedback will influence what is done in future increments. Most iterative approaches also incorporate this feedback loop by delivering some (if not all) of the (intermediate) products created by the iterations.

The painting analogy shown above is not a perfect representation for the iterative approach. If the final product were to comprise 1,000 source code modules, you could be forgiven for thinking that an iterative approach would have people starting the first iteration by writing one line of code in each module and then have the second and subsequent iterations each adding another line of code to each module until they were completed. This is not the case.

In both iterative and incremental models, the features to be implemented are grouped together (for example, according to business priority or risk). In this way, the focus is always on the most important of the outstanding features. The various project phases, including their work products and activities, then occur for each group of features. The phases may be done either sequentially or overlapping, and the iterations or increments themselves may be sequential or overlapping.

An iterative development model is shown in Figure 2.3.

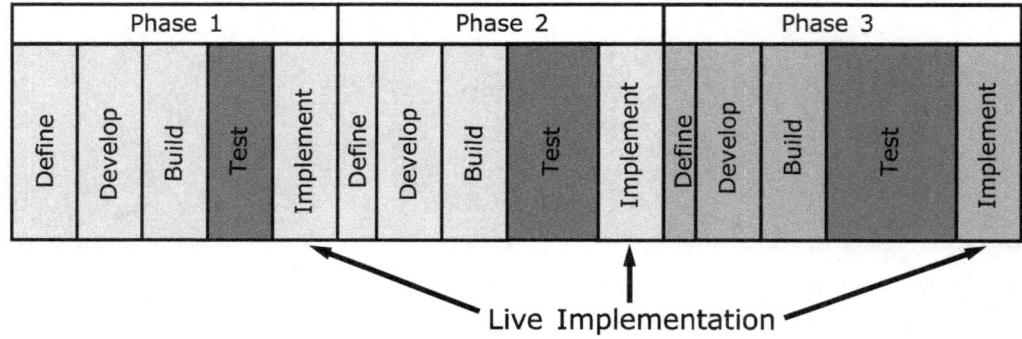

FIGURE 2.3   Iterative development model

Testing in incremental and iterative development

During project initiation, high-level test planning and test analysis occur in parallel with the project planning and business/requirements analysis. Any detailed test planning, test analysis, test design and test implementation occurs at the beginning of each iteration.

Test execution often involves overlapping test levels. Each test level begins as early as possible and may continue after subsequent, higher test levels have started.

In an iterative or incremental life cycle, many of the same tasks will be performed but their timing and extent may vary. For example, rather than being able to implement the entire test environment at the beginning of the project, it may be more efficient to implement only the part needed for the current iteration. The testing tasks may be undertaken in a different order and not necessarily sequentially. There are likely to be fewer entry and exit criteria between activities compared with sequential models. Also, much of the test planning and completion reporting are more likely to occur at the start and end of the project, respectively, rather than at the start and end of each iteration.

With any of the iterative or incremental life cycle models, the further ahead the planning occurs, the further ahead the scope of the test process can extend.

Common issues with iterative and incremental models include:

● more regression testing

● defects outside the scope of the iteration or increment

● less thorough testing.

Because the system is being produced a bit at a time, at any given point there will be some part which is completed in some sense, either an increment or the work that was done iteratively. This part will be tested and may be used by the customer or user to give feedback. When the next increment or iteration is developed, this will also be tested, but it is also important to do regression testing of the parts which have already been developed. The more iterations or increments there are, the more regression testing will be needed throughout development. (This type of testing is a good candidate for automation.)

Defects that are found in the part that you are currently testing are dealt with in the usual way, but what about defects found either by regression testing of previously developed parts, or discovered by accident when testing a new part? These defects do need to be logged and dealt with, but because they are outside the scope of the current iteration/increment, they can sometimes fall between the cracks and be forgotten, neglected or argued about.

There is a danger that the testing may be less thorough in incremental and iterative life cycles, particularly regression testing of previously developed parts, and especially if the regression testing is manual rather than automated. There is also a danger that the testing is less formal, because we are dealing with smaller parts of the system, and formality of the testing may seem like overkill for such a small thing.

Examples of iterative and incremental development models are Rational Unified Process (RUP), Scrum, Kanban and Spiral (or prototyping).

Unified Process (UP)

Unified Process is an incremental software development process framework, originally from Rational Software Development, a division of IBM (and called RUP). UP consists of four steps:

- Inception: the initial idea, planning (for example, what resources would be needed) and go/no-go decision for development, done with stakeholders.
- Elaboration: further detailed investigation into resources, architecture and costs.
- Construction: the software product is developed, including testing.
- Transition: the software product is released to customers, with modifications based on user feedback.

This development process is tailorable for different contexts, and there are tools (and services) to support it. One of the basic principles is that development is iterative, with risk being the primary driver for decisions about the development. Another principle (relevant to us) is that the evaluation of quality (including testing) is continuous throughout development.

In UP, the increments that are produced, although significantly smaller than what is produced by sequential models, are larger than the increments produced by Agile development (refer to Scrum later in the section) and would typically take months rather than days or weeks to complete. They might contain groups of related features, for example.

Spiral (or prototyping)

The Spiral model is an iterative development model based on risk and was initially proposed by Boehm [1996]. There are four steps: determine objectives, identify risks and alternatives, develop and test, and plan the next iteration. Prototypes may be developed as a way of addressing risks. These prototypes may be kept and incorporated into later cycles (an incremental approach), they might be discarded (a throw-away prototype), or they may be reworked as part of the next cycle.

The diagram of the Spiral model shows development starting small from a centre and moving in a circular way clockwise through the four stages. Each succeeding cycle builds outwards in a spiral through the phases, developing more functionality each time. The key driver for the Spiral model is that it is risk-driven.

Agile development

In this section, we will describe what Agile development is and then also cover the changes that this way of working brings to testing.

Agile software development is a group of software development methodologies based on iterative and incremental development, where requirements and solutions evolve through collaboration between self-organizing cross-functional teams. Most Agile teams use Scrum, as described below. Typical Agile teams are five to nine people, and the Agile manifesto describes ways of working that are ideal for small teams and counteract problems prevalent in the late 1990s, with its emphasis on process and documentation. The Agile manifesto consists of four statements describing what is valued in this way of working:

● individuals and interactions over processes and tools

● working software over comprehensive documentation

● customer collaboration over contract negotiation

● responding to change over following a plan.

While there are several Agile methodologies in practice, the industry seems to have settled on the use of Scrum as an Agile management approach, and Extreme Programming (XP) as the main source of Agile development ideas. Some characteristics of project teams using Scrum and XP are:

● The generation of business stories (a form of lightweight use cases) to define the functionality, rather than highly detailed requirements specifications.

● The incorporation of business representatives into the development process, as part of each iteration (called a sprint and typically lasting two to four weeks), providing continual feedback and to define and carry out functional acceptance testing.

● The recognition that we cannot know the future, so changes to requirements are welcomed throughout the development process, as this approach can produce a product that better meets the stakeholders' needs as their knowledge grows over time.

● The concept of shared code ownership among the developers, and the close inclusion of testers in the sprint teams.

● The writing of tests as the first step in the development of a component, and the automation of those tests before any code is written. The component is complete when it then passes the automated tests. This is known as test-driven development.

● Simplicity, by building only what is necessary, not everything you can think of.

● The continuous integration and testing of the code throughout the sprint, at least once a day.

Proponents of the Scrum and XP approaches emphasize testing throughout the process. Each iteration (sprint) culminates in a short period of testing, often with an independent tester as well as a business representative. Developers are to write and run test cases for their code, and leading practitioners use tools to automate those tests and to measure structural coverage of the tests (refer to Chapters 4 and 6). Every time a change is made in the code, the component is tested and then integrated with the existing code, which is then tested using the full set of automated component test cases. This gives continuous integration, by which we mean that changes are incorporated continuously into the software build.

Agile development provides both benefits and challenges for testers. Some of the benefits are:

- The focus on working software and good quality code.
- The inclusion of testing as part of and the starting point of software development (test-driven development).
- Accessibility of business stakeholders to help testers resolve questions about expected behaviour of the system.
- Self-organizing teams, where the whole team is responsible for quality and gives testers more autonomy in their work.
- Simplicity of design that should be easier to test.

There are also some significant challenges for testers when moving to an Agile development approach:

- Testers who are used to working with well-documented requirements will be designing tests from a different kind of test basis: less formal and subject to change. The manifesto does not say that documentation is no longer necessary or that it has no value, but it is often interpreted that way.
- Because developers are doing more component testing, there may be a perception that testers are not needed. But component testing and confirmation-based acceptance testing by only business representatives may miss major problems. System testing, with its wider perspective and emphasis on non-functional testing as well as end-to-end functional testing is needed, even if it does not fit comfortably into a sprint.
- The tester's role is different since there is less documentation and more personal interaction within an Agile team, so testers need to adapt to this style of working, and this can be difficult for some testers. Testers may be acting more as coaches in testing to both stakeholders and developers, who may not have a lot of testing knowledge.
- Although there is less to test in one iteration than a whole system, there is also a constant time pressure and less time to think about the testing for the new features.
- Because each increment is adding to an existing working system, regression testing becomes extremely important, and automation becomes more beneficial. However, simply taking existing automated component or component integration tests may not make an adequate regression suite.

Software engineering teams are still learning how to apply Agile approaches. Agile approaches cannot be applied to all projects or products, and some testing challenges remain to be surmounted with respect to Agile development. However, Agile methodologies are showing promising results in terms of both development efficiency and quality of the delivered code.

More information about testing in Agile development and iterative incremental models can be found in books by Black [2017], Crispin and Gregory [2008] and Gregory and Crispin [2015]. There is an ISTQB certificate called Foundation Level Agile Tester which is for people who have taken a previous version of the Certified

Tester Foundation Level (CTFL) qualification. This current Syllabus, 4.0 (2023), incorporates parts of the previous Agile extension to the foundation level. In order to take the ISTQB Advance Level Technical Agile Tester, you need to have passed either a previous CTFL plus the Foundation Agile Tester, or just this current CTFL.

Scrum

Scrum is an iterative and incremental framework for effective team collaboration, which is typically used in Agile development, the most well-known iterative method. It (and Agile) is based on recognizing that change is inevitable and taking a practical empirical approach to changing priorities. Work is broken down into small units that can be completed in a fairly short time (days, a week or two or even a month). The delivery of a unit of work is called a sprint. For software development, a sprint includes all aspects of development for a particular feature or set of small features, everything from requirements (typically user stories) through to testing and (ideally) test automation.

The development teams are small (five to nine people) and cross-functional, that is, they include people who perform various roles, and often individuals take on different tasks (such as testing) within the team. The key roles are:

● The Product Owner represents the business, that is, stakeholders and end users.

● The Development team (which includes testers) makes its own decisions about development, that is, they are self-organizing with a high level of autonomy.

● The Scrum Master helps the development team to do their work as efficiently as possible, by interacting with other parts of the organization and dealing with problems. The Scrum Master is not a manager, but a facilitator for the team.

A stand-up meeting of typically around 15 minutes is held each day, for example first thing in the morning, to update everyone with progress from the previous day and plan the work ahead. (This is where the term 'scrum' came from, as in the gathering of a rugby team.)

At the start of a sprint, in the sprint planning meeting, some features are selected to be implemented, with other features being put on a backlog. Acceptance criteria apply to user stories and are similar to test conditions, saying what needs to work for the user story to be considered working. A definition of done can apply to a user story (which includes but goes beyond satisfaction of the acceptance criteria), but also to unit testing, system testing, iterations and releases. After a sprint completes, a retrospective should be held to assess what went well and what could be improved for the next sprint.

Because development is limited to the sprint duration which is time-boxed, flexibility is in choosing what can be developed in the time. Compare that to sequential models, where all the features are selected first, and the time taken to develop all of them is based on that. Thus Scrum (and Agile) enable us to deliver the greatest value soonest, an approach first proposed in the 1980s.

Because the iterations are short, the increments are small, such as a few small features or even a few enhancements or bug fixes.

Kanban

Kanban came from an approach to work in manufacturing at Toyota. It is a way of visualizing work and workflow. A Kanban board has columns for different stages of work, from initial idea through development and testing stages to final delivery to users. The tasks are put on sticky notes which are moved from left to right through the columns (like an assembly line for cars).

A key principle of Kanban is to have a limit for work-in-progress activities. If we concentrate on one task, we are much more efficient at doing it, so this approach is less wasteful than trying to do little bits of lots of different tasks. This focus on eliminating waste makes this a lean approach.

There is also a strong focus on user and customer needs. Iterations can be a fixed length to deliver a single feature or enhancement, or features can be grouped together for delivery. Kanban can span more than one team's work (as opposed to Scrum). If user stories are grouped by feature, work may span more than one column on the Kanban board, sometimes referred to as swim lanes.

2.1.1 Impact of the software development lifecycle on testing

The life cycle model that is adopted for a project will have a big impact on the testing that is carried out. Testing does not exist in isolation; test activities are highly related to software development activities. It will define the what, where and when of our planned testing, influence regression testing and largely determine which test techniques to use. The way testing is organized must fit the development life cycle or it will fail to deliver its benefit. If time to market is the key driver, then the testing must be fast and efficient. If a fully documented software development life cycle, with an audit trail of evidence, is required, the testing must be fully documented.

The choice of the software lifecyle model typically impacts the following:

- Scope and timing of test activities, test levels and test types.
- Level of detail of test documentation.
- Choice of test techniques and test approach.
- Extent of test automation.
- Role and responsibility of a tester.

In sequential development models, the tester will often participate in the requirements view early in the lifecycle. Once the requirements have been agreed upon, the system will be designed and subsequently the code will be developed. Only after that will dynamic testing commence, starting with component testing, then component integration testing, system testing and finally user acceptance testing. Sequential development models often coincide with more detailed test documentation and an independent test organization that performs system and/or user acceptance testing.

Typically, in iterative and incremental development models, each iteration delivers a working prototype of an increment. This implies that in each iteration, static and dynamic testing are performed. Frequent delivery of increments also means fast user feedback and thus early validation of the product being delivered. Since we need to ensure that software code from previous iterations is still behaving as intended, extensive regression testing is needed and therefore automation is more important.

In Agile software development change is embraced; it often happens in the course of a project. To deal with change, (test) documentation is typically lightweight and flexible, with experience-based test techniques, for example exploratory testing, being used. Experience-based test techniques (refer to Section 4.4) require less extensive test analysis and design. To manage the risk of change, regression testing is important, which typically also means test automation is introduced to make regression testing more efficient.

2.1.2 Software development lifecycle and good testing practices

Whichever life cycle model is being used, there are several characteristics of good testing:

● For every development activity there is a corresponding test activity, so that all development activities are subject to quality control.

● Each test level (refer to Section 2.2) has test objectives specific to that level, which allows for testing to be appropriately comprehensive while avoiding redundancy.

● The analysis and design of tests for a given test level should begin during the corresponding software development activity.

● Testers should participate in discussions to help define and refine requirements and design. They should also be involved in reviewing work products as soon as drafts are available in the software development cycle.

Recall from Chapter 1, testing Principle 3: 'Early testing saves time and money'. By starting testing activities as early as possible in the software development life cycle, we find defects while they are still small green shoots (in requirements, for example) before they have had a chance to grow into trees (in production). We also prevent defects from occurring at all by being more aware of what should be tested from the earliest point in whatever software development life cycle we are using. Note that this earlier testing and defect detection also support the shift-left approach (refer to Section 2.1.5)

2.1.3 Testing as a driver for software development

Test-Driven Development (TDD), Acceptance Test-Driven Development (ATDD) and Behaviour-Driven Development (BDD) are approaches often used in an Agile context, where tests are defined also as a means of directing development. These approaches are an implementation of early testing (refer to Section 1.3) and also follow the shift-left approach (refer to Section 2.1.5) since the tests are defined before the code is written.

Test-Driven Development (TDD)

Test-Driven Development (TDD) is a software development approach in which tests are written before the code that needs to be implemented. As such, TDD directs the coding through test cases instead of extensive software design.

The TDD process typically follows these steps:

1 Write a test: Before writing any code, a developer writes a test that defines a function or improvements to a function, which should fail because the function is not yet implemented.

2 Run the test: The newly written test is executed to make sure it fails. This step confirms that the test is correctly detecting the absence of the expected functionality.

3 Write code: The developer writes the minimum amount of code necessary to pass the test. The focus is on making the test pass, not on writing the entire functionality.

4 Run all tests: After implementing the code, all existing tests (not just the new one) are run to ensure that the new code did not break any existing functionality.

5 Refactor code: If the code meets the requirements and all tests pass, the developer may refactor the code for better design or performance while ensuring that the tests continue to pass.

6 Repeat: Steps 1–5 are repeated for each new piece of functionality.

TDD relies heavily on automated testing to quickly run and verify the correctness of the code. It promotes small, incremental changes to the codebase, making it easier to identify and fix issues early in the development process.

Acceptance Test-Driven Development (ATDD)

With Acceptance Test-Driven Development (ATDD) (refer also to Section 4.5.3) software development is driven by a set of acceptance criteria or tests that define the desired behaviour of the system from the perspective of the end user. Agile software development is often organized around user stories, which are short, simple descriptions of a feature or functionality from an end user's perspective. Each user story is accompanied by acceptance criteria, which are detailed conditions that must be satisfied for the story to be considered complete. In ATDD acceptance tests are written based on the acceptance criteria defined for each user story. These tests serve as executable specifications and are typically written in a language that is readable by both technical and non-technical team members. While acceptance tests can be written manually, ATDD often encourages the automation of these tests to ensure they can be easily and repeatedly executed. Automated tests help catch regressions and ensure that new changes do not break existing functionality.

By following the principles of ATDD, development teams aim to create software that not only meets technical requirements but also fulfils the business goals and expectations. This approach helps in building a shared understanding of the desired functionality and promotes a more transparent and collaborative development process.

Behaviour-Driven Development (BDD)

Behaviour-Driven Development (BDD) extends the principles of TDD to include collaboration between developers, testers and non-technical stakeholders such as business analysts and product owners. BDD focuses on describing the expected behaviour of a system through scenarios written in natural language, making it more accessible to non-technical team members.

BDD typically starts with the creation of user stories. Each user story is accompanied by one or more scenarios that describe specific examples of how the feature should behave under different conditions. Scenarios in BDD are often written in a specific format called Gherkin, which uses a Given-When-Then structure. Gherkin serves as a domain-specific language for expressing the behaviour of a system in a way that is easily understandable by both technical and non-technical stakeholders. The Given-When-Then structure helps in organizing the scenario into three parts:

1 Given: Describes the initial state or preconditions of the system.

2 When: Describes the specific action or event that triggers the behaviour.

3 Then: Describes the expected outcome or result of the behaviour.

For example:

Scenario: Successful login

- given the user is on the login page
- when the user enters valid credentials
- then the user should be logged in successfully.

This example scenario describes the behaviour of a login functionality in a way that is clear and understandable by both technical and non-technical team members. The actual implementation of the steps (Given, When, Then) in code can be done using one of the automation frameworks that support BDD.

2.1.4 DevOps and testing

DevOps is a set of practices that aims to automate and improve the collaboration and communication between software development (Dev) and IT operations (Ops) teams. The goal of DevOps is to shorten the system's development life cycle and deliver high-quality software continuously through a DevOps delivery pipeline. Key principles and practices of DevOps include:

- Collaboration: DevOps emphasizes collaboration and communication between development and operations teams. This helps in breaking down silos and creating a more integrated and efficient workflow.
- Automation: Automation is a critical aspect of DevOps. It involves using tools to automate repetitive tasks, allowing teams to focus on more strategic and creative work. Automation helps in reducing errors and increasing the speed of software delivery.
- Continuous Integration (CI): CI is a development practice where developers integrate their code into a shared repository frequently, typically multiple times a day. Each integration triggers automated tests to ensure that the new code does not introduce bugs.
- Continuous Delivery (CD): CD is an extension of continuous integration where the software is semi-automatically deployed to production environments after passing automated tests. This enables more frequent and reliable software releases.
- Feedback Loops: DevOps encourages the creation of feedback loops throughout the development and operations processes. Feedback helps teams to learn from their experiences and continuously improve their processes.

DevOps practices enhance the testing process by promoting automation, collaboration and continuous testing throughout the software development life cycle. From a testing perspective, the key benefits of DevOps are:

- DevOps practices, including continuous monitoring and feedback loops, provide valuable information to testing teams. Fast feedback is provided on code quality and whether changes adversely affect existing code.
- DevOps encourages a 'shift-left' approach to testing, meaning that testing activities are moved earlier in the development process. This shift-left strategy ensures that potential issues are identified and addressed as early as possible, reducing the cost of fixing defects later in the cycle. The shift-left strategy within DevOps includes encouraging developers to put more emphasis on component testing and static analysis.

- With the adoption of automated processes like CI/CD testing is extended beyond application code to test environment provisioning and configuration. DevOps ensures that the test environment is more reliable, scalable and consistent.

- There is more focus on non-functional quality characteristics such as performance and reliability.

- DevOps promotes the automation of various testing processes, including component testing, integration testing and system testing. Automated testing ensures consistency, repeatability and faster feedback, allowing teams to release software more confidently and frequently. It reduces the need for repetitive manual testing, although some manual testing will still be needed especially from the user's perspective.

- Automation allows for more extensive regression testing. Automated tests can be executed more frequently and consistently than manual tests, which may lead to more comprehensive coverage of the application's functionality thereby reducing the risk in regression.

Of course, the implementation of DevOps is not without its challenges. The DevOps delivery pipeline must be defined and established, CI/CD tools must be introduced and maintained, and test automation requires additional knowledge, skills and resources, and may be difficult to establish and maintain.

2.1.5 Shift-left approach

The term shift-left in testing refers to the practice of moving testing activities earlier in the software development lifecycle (SDLC) and adheres to the principle of early testing (Section 1.3). Traditionally, testing is often performed after the development phase, but the shift-left approach advocates for conducting testing activities earlier, sometimes even before the actual coding begins. The shift-left approach is aligned with Agile and DevOps methodologies, where there is a focus on delivering software quickly and continuously improving it based on feedback. By identifying and addressing issues early in the development process, the shift-left approach aims to contribute to the overall improvement of software quality and helps in delivering more reliable and robust software systems.

A **shift-left** approach will result in additional effort earlier in the lifecycle but is thus expected to save effort later in the lifecycle. As investments are required early in the lifecyle sometimes giving the perception the process is slowing down, it is important that stakeholders understand, are convinced and are brought into the shift-left concept. Testing is often thinking of what could go wrong; if you think of what mistakes you might make before you start writing the code, you are much less likely to make those mistakes.

> **Shift-left** An approach to performing testing and quality assurance activities as early as possible in the software development lifecycle.

Examples of good practices that illustrate how to achieve shift-left in testing include:

- Reviewing, which involves conducting various types of reviews and inspections early in the software development lifecycle to catch issues, improve collaboration and enhance overall product quality. For example, requirements reviews will identify ambiguities, address inconsistencies and ensure a shared understanding of the requirements.

- Designing and writing test cases before the code is written. Early test design will typically identify ambiguities in the test basis documentation. Other examples include TDD, ATDD and BDD (refer to Section 2.1.3). As discussed earlier in TDD, developers create test cases that define the expected behaviour of a feature and then write code to make those tests pass.

- Continuous testing through CI and CD, where testing is integrated into the continuous integration and continuous delivery (CI/CD) pipeline, ensuring fast feedback. With this practice, testing is an ongoing and integral part of the development process. At component level, automated component tests accompany source code when it is submitted to the code repository.

- Completing static analysis on source code prior to dynamic testing. Static code analysis is a type of static analysis that specifically focuses on examining the source code of a program without executing it. The primary goal of static code analysis is to find issues, defects or vulnerabilities in the code before the software is run. This analysis is typically automated and is often performed using specialized tools.

- Performing non-functional testing as early as possible. Most often non-functional testing, for example performance testing, reliability testing, security testing, is performed at a late stage in the SDLC when a complete system and a representative test environment are available. However, most often there are possibilities to start much earlier, for example at component test level, addressing these non-functional aspects of a system. For example, performance testing tools can be applied to analyze code and infrastructure performance as part of the CI/CD pipeline and static analysis tools can identify potential security vulnerabilities in the code, such as SQL injection, cross-site scripting and other common issues.

2.1.6 Retrospectives and process improvement

In more mature organizations, both sequential and iterative life cycles will typically include post-implementation reviews, lessons learned meetings and other opportunities to gather feedback and implement improvements. All of these methods and techniques are typically covered under the general heading of retrospective meeting. With sequential life cycles applied to longer projects, the retrospective meeting may also take place at the end of each milestone and not just at the end of the project.

Retrospective meetings are a great chance to recap how the work went and what lessons were learned. In general, a project retrospective helps to optimize teamwork and communication and creates more fun and better results during a project. By means of a retrospective, the team gains new insights into what went well and what to do better. At the end there are always lessons to be learned.

A retrospective is an open discussion. The sessions also need to be a blame-free psychologically safe space for which the retrospective facilitator's role is critical. Attendees are invited to share thoughts and experiences and come up with lessons learned. The team starts talking about its project experience by using standard questions like these:

- What was successful and should be retained?
- What was not so successful and could be improved?
- How can we incorporate the improvements and retain the successes in the future?

It is usually surprising how much you can learn in a retrospective meeting for the next project. The results should be recorded, and test-related experiences and improvement ideas are normally part of the test completion report (refer to Section 5.3.2).

In iterative methodologies with short iterations, (for example, Agile methodologies), the feedback loops will happen more frequently, and therefore the opportunities to implement improvements are more frequent. For example, Agile development life cycle models expect a project retrospective at the end of each iteration. The evaluation should identify and prioritize the major items that went well and those items that, if done differently, could make things even better. By the end of the sprint retrospective, the team should have identified actionable improvement measures that it implements in the next iteration.

Retrospectives are critical to a successful implementation of continuous improvements. Typical benefits for testing include:

- Increased test effectiveness: teams can use retrospectives to identify gaps in their testing processes. This includes areas such as risk analysis, test case design and execution. By addressing these gaps, the testing process becomes more robust.

- Increased test efficiency: retrospectives allow teams to discuss changes in process, technology, tools or testware with a focus on making testing practices more efficient.

- Increased quality of testware: by reviewing test work products and discussing test processes, the testware's content and structure can be improved.

- Team bonding and learning: by acknowledging achievements and discussing challenges in a constructive manner, retrospectives contribute to team morale. Engaged and motivated team members are more likely to deliver high-quality testing outcomes.

- Improved quality of the test basis: retrospectives provide an opportunity to conduct root cause analysis for any issues or defects that were identified during testing. Understanding the root causes helps in implementing preventive measures for future releases. A typical root cause is often the quality level of the requirements.

- Better cooperation between development and testing: retrospectives encourage open communication and collaboration among team members. This can lead to better coordination between developers and testers, often also improving the overall efficiency of the testing efforts.

2.2 TEST LEVELS AND TEST TYPES

SYLLABUS LEARNING OBJECTIVES FOR 2.2 TEST LEVELS AND TEST TYPES (K2)

FL-2.2.1   **Distinguish the different test levels (K2)**

FL-2.2.2   **Distinguish the different test types (K2)**

FL-2.2.3   **Distinguish confirmation testing from regression testing**

We have mentioned (and given the definition for) test levels in Section 2.1. The definition of 'test level' is 'a specific instance of the test process', which is not necessarily the most helpful. The Syllabus here describes test levels as:

'groups of test activities that are organized and managed together'

The test activities were described in Chapter 1, Section 1.4 (test planning through to test completion). When we talk about test levels, we are looking at those activities performed with reference to development levels (such as those described in the V-model) from components to systems, or even systems of systems.

In this section, we will look in more detail at the various test levels and show how they are related to other activities within the software development life cycle. The key characteristics for each test level are discussed and defined, to be able to more clearly separate the various test levels. A thorough understanding and definition of the various test levels will identify missing areas and prevent overlap and repetition. Sometimes we may wish to introduce deliberate overlap to address specific risks. Understanding whether we want overlaps and removing the gaps will make the test levels more complementary, leading to more effective and efficient testing. We will look at four test levels in this section.

As we go through this section, watch for the Syllabus terms **acceptance testing**, **black-box testing**, **component integration testing**, **component testing**, **confirmation testing**, **functional testing**, **integration testing**, **non-functional testing**, **regression testing**, **system integration testing**, **system testing**, **test object**, **test type** and **white-box testing**. These terms are also defined in the Glossary and in the ISTQB online Glossary.

While the specific test levels required for—and planned for—a particular project can vary, good practice in testing suggests that each test level has the following clearly identified:

- Specific test objectives for the test level.
- The test basis, the work product(s) used to derive the test conditions and test cases.
- The **test object** (that is, what is being tested such as an item, build, feature or system under test).
- The typical defects and failures that we are looking for at this test level.
- Specific approaches and responsibilities for this test level.

Test object The work product to be tested.

One additional aspect is that each test level needs a test environment. Sometimes an environment can be shared by more than one test level; in other situations, a particular environment is needed. For example, acceptance testing should have a test environment that is as similar to production as is possible or feasible. In component testing, developers often just use their development environment. In system testing, an environment may be needed with particular external connections, for example.

When these topics are clearly understood and defined for the entire project team, this contributes to the success of the project. In addition, during test planning, the managers responsible for the test levels should consider how they intend to test a system's configuration, if such data is part of a system.

Test levels may overlap in time; in Agile development different iterations may be at different test levels at the same time, in sequential development, or the exit criteria for one test level may be the entry criteria for the next.

2.2.1 Test levels

2.2.1.1 *Component testing*

Component testing, also known as unit or module testing, searches for defects in, and verifies the functioning of, software items (for example, modules, programs, objects, classes, etc.) that are separately, in isolation, testable.

Component tests are typically based on the requirements and detailed design specifications applicable to the component under test, as well as the code itself (which we will discuss in Chapter 4 when we talk about white-box testing).

The component under test, the test object, includes the individual components, the data conversion and migration programs used to enable the new release, and database tables, joins, views, modules, procedures, referential integrity and field constraints, and even whole databases.

> **Component testing**
> (module testing, unit testing) A test level that focuses on individual hardware or software components.

Component testing: objectives

The different test levels have different objectives. The objectives of component testing include:

- Reducing risk (for example, by testing high-risk components more extensively).
- Verifying whether or not functional and non-functional behaviours of the component are as they should be (as designed and specified).
- Building confidence in the quality of the component, which may include measuring structural coverage of the tests, giving confidence that the component has been tested as thoroughly as was planned.
- Finding defects in the component.
- Preventing defects from escaping to later testing.

In incremental and iterative development (for example, Agile), automated component regression tests are run frequently, to give confidence that new additions or changes to a component have not caused existing components or links to break.

Component testing may be done in isolation from the rest of the system depending on the context of the development life cycle and the system. Most often, mock objects or stubs and drivers are used to replace the missing software and simulate the interface between the software components in a simple manner. A stub or mock object is called from the software component to be tested; a driver calls a component to be tested (refer to Figure 2.4). Test harnesses may also be used to provide similar functionality, and service virtualization can give cloud-based functionality to test components in realistic environments.

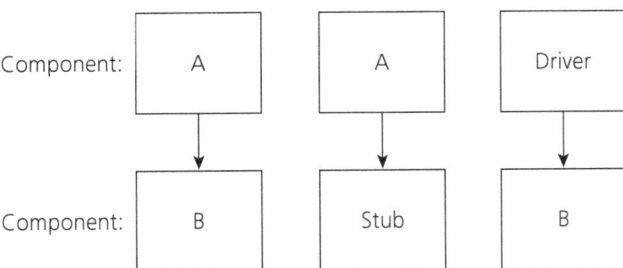

FIGURE 2.4   Stubs and drivers

Component testing may include testing of functionality (for example, are the calculations correct) and specific non-functional characteristics such as resource-behaviour (for example, memory leaks), performance testing (for example, do calculations complete quickly enough), as well as structural testing. Test cases are derived from work products such as the software design or the data model.

Component testing: test basis

What is this particular component supposed to do? Examples of work products that can be used as a test basis for component testing include:

- detailed design
- code
- data model
- component specifications (if available).

Component testing: test objects

What are we actually testing at this level? We could say the smallest thing that can be sensibly tested on its own. Typical test objects for component testing include:

- components themselves, units or modules
- code and data structures
- classes
- database models.

Component testing: typical defects and failures

Examples of defects and failures that can typically be revealed by component testing include:

- incorrect functionality (for example, not as described in a design specification)
- data flow problems
- incorrect code or logic.

At the end of the description of all the test levels, refer to Table 2.1, which summarizes the characteristics of each test level.

Component testing: specific approaches and responsibilities

Typically, component testing occurs with access to the code being tested and with the support of the development environment, such as a unit test framework or debugging tool. In practice it usually involves the developer who wrote the code. The developer may change between writing code and testing it. Sometimes, depending on the applicable level of risk, component testing is carried out by a different developer, introducing independence. Defects are typically fixed as soon as they are found, without formally recording them in a defect management tool. Of course, if such defects are recorded, this can provide useful information for root cause analysis.

One approach in component testing, initially developed in Extreme Programming (XP), is to prepare and automate test cases before coding. This is called a test-first approach or test-driven development (TDD) (refer to Section 2.1.3). This approach is highly iterative and is based on cycles of developing automated tests, then building and integrating small pieces of code, then executing the component tests until they pass; it is typically done in Agile development. The idea is that the first thing the developer does is to write some automated tests for the component. Of course, if these are run now, they will fail because no code is there! Then just enough code is written until those tests pass. This may involve fixing defects now found by the tests and re-factoring the code. (This approach also helps to build only what is needed rather than a lot of functionality that is not really wanted.)

2.2.1.2   Integration testing

The Syllabus lists two types of integration testing as two of the five test levels. Since two of them are a form of integration testing, we will discuss both component and system integration testing in this section, although sequentially, system integration testing comes after system testing.

Integration testing tests interfaces between components and interactions of different parts of a system such as an operating system, file system and hardware or interfaces between systems. Integration tests are typically based on the software and system design (both high-level and low-level), the system architecture (especially the relationships between components or objects) and the workflows or use cases by which the stakeholders will employ the system.

> **Integration testing**
> A test level that focuses on interactions between components or systems.

The two types of integration testing mentioned in the Syllabus are:

- **Component integration testing** (also known as unit integration testing) tests the interactions between software components and is done after component testing. Component integration testing is performed to expose defects in the interfaces and interactions between integrated components. It is a good candidate for automation. In iterative and incremental development, both component tests and integration tests are usually part of continuous integration, which may involve automated build, test and release to end users or to a next level. At least, this is the theory. In practice, component integration testing may not be done at all, or is misunderstood, and as a consequence is not done well.

> **Component integration testing**
> The integration testing of components.

- **System integration testing** tests the interactions between different systems, packages and microservices, and may be done after system testing. It is testing focused on the combination and interaction of systems. System integration testing may also test interfaces to and provided by external organizations (such as web services). In this case, the developing organization may control only one side of the interface, resulting in a number of problems: changes may be destabilizing, defects in the external organization's software may block progress in the testing, or special test environments may be needed. Business processes implemented as workflows may involve a series of systems that can even run on different platforms. System integration testing may be done in parallel with other testing activities. The test environment for system integration testing should be (as much as possible) similar to the operational environment.

> **System integration testing** The integration testing of systems.

Integration testing: objectives

The objectives of integration testing at either component or system level include:

- Reducing risk, for example by testing high-risk integrations first.
- Verifying whether or not functional and non-functional behaviours of the interfaces are as they should be, as designed and specified.
- Building confidence in the quality of the interfaces.
- Finding defects in the interfaces themselves or in the components or systems being tested together.
- Preventing defects from escaping to later testing.

Automated integration regression tests (such as in continuous integration) provide confidence that changes have not broken existing interfaces, components or systems.

Integration testing: test basis

How are these components or systems supposed to work together and communicate? Examples of work products that can be used as a test basis for integration testing include:

- software and system design
- sequence diagrams
- interface and communication protocol specifications
- use cases
- architecture at component or system level
- workflows
- external interface definitions.

Integration testing: test objects

What are we actually testing at this level? The emphasis here is in testing things with others which have already been tested individually. We are interested in how things work together and how they interact. Typical test objects for integration testing include:

- subsystems
- databases
- infrastructure
- interfaces
- APIs (Application Programming Interfaces)
- microservices.

Integration testing: typical defects and failures

Examples of defects and failures that can typically be revealed by component integration testing include:

- Incorrect data, missing data or incorrect data encoding.
- Incorrect sequencing or timing of interface calls.
- Interface mismatch, for example where one side sends a parameter where the value exceeds 1,000, but the other side only expects values up to 1,000.
- Failures in communication between components.
- Unhandled or improperly handled communication failures between components.
- Incorrect assumptions about the meaning, units or boundaries of the data being passed between components.

Examples of defects and failures that can typically be revealed by system integration testing include:

- Inconsistent message structures between systems.
- Incorrect data, missing data or incorrect data encoding.

- Interface mismatch.
- Failures in communication between systems.
- Unhandled or improperly handled communication failures between systems.
- Incorrect assumptions about the meaning, units or boundaries of the data being passed between systems.
- Failure to comply with mandatory security regulations.

At the end of the description of all the test levels, refer to Table 2.1, which summarizes the characteristics of each test level.

Integration testing: specific approaches and responsibilities

The greater the scope of integration, the more difficult it becomes to isolate failures to a specific interface, which may lead to an increased risk. This leads to varying approaches to integration testing. One extreme is that all components or systems are integrated simultaneously, after which everything is tested as a whole. This is called big-bang integration. Big-bang integration has the advantage that everything is finished before integration testing starts. There is no need to simulate (as yet unfinished) parts. The major disadvantage is that in general it is time-consuming and difficult to trace the cause of failures with this late integration. So big-bang integration may seem like a good idea when planning the project, being optimistic and expecting to find no problems. If you think integration testing will find defects, it is a good practice to consider whether time might be saved by breaking down the integration test process.

Another extreme is that all programs are integrated one by one, and tests are carried out after each step (incremental testing). Between these two extremes, there is a range of variants. The incremental approach has the advantage that the defects are found early in a smaller assembly when it is relatively easy to detect the cause. A disadvantage is that it can be time-consuming since mock objects or stubs and drivers may have to be developed and used in the test. Within incremental integration testing a range of possibilities exist, partly depending on the system architecture:

- Top-down: testing starts from the top and works to the bottom, following the control flow or architectural structure (for example, starting from the graphical user interface (GUI) or main menu). Components or systems are substituted by stubs.
- Bottom-up: testing reverses this approach, starting from the bottom of the control flow upwards. Components or systems are substituted by drivers.
- Functional incremental: integration and testing take place on the basis of the functions or functionality, as documented in the functional specification.

The preferred integration sequence and the number of integration steps required depend on the location in the architecture of the high-risk interfaces. The best choice is to start integration with those interfaces that are expected to cause the most problems. Doing so prevents major defects at the end of the integration test stage. In order to reduce the risk of late defect discovery, integration should normally be incremental rather than big bang. Ideally, testers should understand the architecture and influence integration planning. If integration tests are planned before components or systems are built, they can be developed in the order required for most efficient testing. A risk analysis of the most complex interfaces can help to focus integration testing. In iterative and incremental development, integration is also incremental. Existing integration tests should be part of the regression tests used in continuous integration. Continuous integration has major benefits because of its iterative nature.

At each stage of integration, testers concentrate solely on the integration itself. For example, if they are integrating component A with component B, they are interested in testing the communication between the components, not the functionality of either one. In integrating system X with system Y, again the focus is on the communication between the systems and what can be done by both systems together, rather than defects in the individual systems. Both functional and structural approaches may be used. Testing of specific non-functional characteristics (for example, performance) may also be included in integration testing.

Component integration testing is often carried out by developers; system integration testing is generally the responsibility of the testers. Either type of integration testing could be done by a separate team of specialist integration testers, or by a specialist group of developers/integrators, including non-functional specialists. The testers performing the system integration testing need to understand the system architecture. Ideally, they should have had an influence on the development, integration planning and integration testing.

2.2.1.3 System testing

System testing A test level that focuses on verifying that a system as a whole meets specified requirements.
(Note that the ISTQB definition implies that system testing is only about verification of specified requirements. In practice, system testing is often also about validation that the system is suitable for its intended users, as well as verifying against any type of requirement.)

System testing is concerned with the behaviour of the whole system/product as defined by the scope of a development project or product. It may include tests based on risk analysis reports, system, functional or software requirements specifications, business processes, use cases or other high-level descriptions of system behaviour, interactions with the operating system and system resources. The focus is on end-to-end tasks that the system should perform, including non-functional aspects, such as performance.

In some systems, the quality of the data may be of critical importance, so there would be a focus on data quality. System level tests may be automated to provide a regression suite to ensure that changes have not adversely affected existing system functionality. Stakeholders may use the information from system testing to decide whether the system is ready for user acceptance testing, for example. System testing is also where conformance to legal or regulatory requirements or to external standards is tested.

The test environment is important for system testing; it should correspond to the final production environment as much as possible.

System testing: objectives

The objectives of system testing include:

- reducing risk
- verifying whether or not functional and non-functional behaviours of the system are as they should be (as specified)
- validating that the system is complete and will work as it should and as expected
- building confidence in the quality of the system as a whole
- finding defects
- preventing defects from escaping to later testing or to production.

System testing: test basis

What should the system as a whole be able to do? Examples of work products that can be used as a test basis for system testing include:

- software and system requirement specifications (functional and non-functional)
- risk analysis reports
- use cases
- epics and user stories
- models of system behaviour
- state diagrams
- system and user manuals.

System testing: test objects

What are we actually testing at this level? The emphasis here is in testing the whole system, from end to end, encompassing everything that the system needs to do (and how well it should do it, so non-functional aspects are also tested here). Typical test objects for system testing include:

- applications
- hardware/software systems
- operating systems
- system under test (SUT)
- system configuration and configuration data.

System testing: typical defects and failures

Examples of defects and failures that can typically be revealed by system testing include:

- incorrect calculations
- incorrect or unexpected system functional or non-functional behaviour
- incorrect control and/or data flows within the system
- failure to properly and completely carry out end-to-end functional tasks
- failure of the system to work properly in the production environment(s)
- failure of the system to work as described in system and user manuals.

At the end of the description of all the test levels, refer to Table 2.1, which summarizes the characteristics of each test level.

System testing: specific approaches and responsibilities

System testing is most often the final test on behalf of development to verify that the system to be delivered meets the specification and to validate that it meets expectations; one of its purposes is to find as many defects as possible. Most often it is carried out by specialist testers that form a dedicated, and sometimes independent, test team within development, reporting to the development manager or project manager. In some organizations system testing is carried out by a third-party team or by business analysts. Again, the required level of independence is based on the applicable risk level, and this will have a high influence on the way system testing is organized.

System testing should investigate end-to-end behaviour of both functional and non-functional aspects of the system. An end-to-end test may include all of the steps in a typical transaction, from logging on, accessing data, placing an order, etc. through to logging off and checking order status in a database. Typical non-functional tests include performance, security and reliability. Testers may also need to deal with incomplete or undocumented requirements. System testing of functional requirements starts by using the most appropriate black-box techniques for the aspect of the system to be tested. For example, a decision table may be created for combinations of effects described in business rules. White-box techniques may also be used to assess the thoroughness of testing elements such as menu dialogue structure or web page navigation (refer to Chapter 4 for more on test techniques).

System testing requires a controlled test environment with regard to, among other things, control of the software versions, testware and the test data (refer to Chapter 5 for more on configuration management). A system test is executed by the development organization in a (properly controlled) environment. The test environment should correspond to the final target or production environment as much as possible in order to minimize the risk of environment-specific failures not being found by testing.

System testing is often carried out by independent testers, for example an internal test team or external testing specialists. However, if testers are only brought in when system test execution is about to start, you will miss a lot of opportunities to save time and money, as well as aggravation. If there are defects in specifications, such as missing functions or incorrect descriptions of business processes, these may not be picked up before the system is built. Because many defects result from misunderstandings, the discussions (indeed arguments) about them tend to be worse the later they are discovered. The developers will defend their understanding because that is what they have built. The independent testers or end-users may realize that what has been built was not what was wanted. This situation can lead to defects being missed in testing (if they are based on wrong specifications) or things being reported as defects that actually are not (due to misunderstandings). These are known as false negative and false positive results respectively. Referring back to testing Principle 3, early test involvement saves time and money, so have testers involved in user story refinement and static testing such as reviews.

2.2.1.4 Acceptance testing

Acceptance testing
A test level that focuses on determining whether to accept the system.

When the development organization has performed its system test (and possibly also system integration tests) and has corrected all or most defects, the system may be delivered for **acceptance testing**. Acceptance tests typically produce information to assess the system's readiness for release or deployment to end-users or customers. Although defects are found at this level, that is not the main aim of acceptance testing. (If lots of defects are found at this late stage, there are serious problems with the whole system, and major project risks.) The focus is on validation, the use of the system for real, how suitable the system is to be put into production or actual use by its intended users. Regulatory and legal requirements, and conformance to standards may also be checked in acceptance testing, although they should also have been addressed in an earlier level of testing, so that the acceptance test is confirming compliance to the standards.

Acceptance testing: objectives

The objectives of acceptance testing include:

- Establishing confidence in the quality of the system as a whole.
- Validating that the system is complete and will work as expected.
- Verifying that functional and non-functional behaviours of the system are as specified.

Different forms of acceptance testing

Acceptance testing is quite a broad category, and it comes in several different flavours or forms. We will look at four of these.

User acceptance testing (UAT)

User acceptance testing is exactly what it says. It is acceptance testing done by (or on behalf of) users, that is, end-users. The focus is on making sure that the system is really fit for purpose and ready to be used by real intended users of the system. The UAT can be done in the real environment or in a simulated operational environment (but as realistic as possible). The aim of testing here is to build confidence that the system will indeed enable the users to do what they need to do in an efficient way. The system needs to fulfil the requirements and meet their needs. The users focus on their business processes, which they should be able to perform with a minimum of difficulty, cost and risk.

Operational acceptance testing (OAT)

Operational acceptance testing focuses on operations and may be performed by system administrators. The main purpose is to give confidence to the system administrators or operators that they will be able to keep the system running and recover from adverse events quickly and without additional risk. It is normally performed in a simulated production environment and is looking at operational aspects, such as:

- testing of backups and restoration of backups
- installing, uninstalling and upgrading
- disaster recovery
- user management
- maintenance tasks
- data loading and migration tasks
- checking for security vulnerabilities (for example, ethical hacking)
- performance and load testing.

Contractual and regulatory acceptance testing

If a system has been custom-developed for another company, there is normally a legal contract describing the responsibilities, schedule and costs of the project. The contract should also include or refer to acceptance criteria for the system, which should have been defined and agreed when the contract was first taken out. Having agreed the acceptance criteria in advance, contractual acceptance testing is focused on whether or not the system meets those criteria. This form of testing is often performed by users or independent testers.

Regulatory acceptance testing is focused on ensuring that the system conforms to government, legal or safety regulations. This type of testing is also often performed by independent testers. It may be a requirement to have a representative of the regulatory body present to witness or to audit the tests.

For both of these forms of acceptance testing, the aim is to build confidence that the system is in conformance with the contract or regulations.

Alpha and beta testing

Alpha and beta testing are typically used for commercial off-the-shelf (COTS) software, such as software packages that can be bought or downloaded by consumers. Feedback is needed from potential or existing users in their market before the software product is put out for sale commercially. The testing here is looking for feedback (and defects) from real users and customers or potential customers. Sometimes free software is offered to those who volunteer to do beta testing.

The difference between alpha and beta testing is only in where the testing takes place. Alpha testing is at the company that developed the software, and beta testing is done in the users' own offices or homes. In alpha testing, a cross-section of potential users are invited to use the system. Developers observe the users and note problems. Alpha testing may also be carried out by an independent test team. Alpha testing is normally mentioned first, but these two forms can be done in any order, or only one could be done (or none).

Beta testing sends the system or software package out to a cross-section of users who install it and use it under real-world working conditions. The users send records of defects with the system to the development organization, where the defects are repaired. Beta testing is more visible and is increasingly popular to be done remotely. For example, crowd testing, where people or potential users from all over the world remotely test an application, can be a form of beta testing. One of the advantages of beta testing is that different users will have a great variety of different environments (browsers, other software, hardware configurations, etc.), so the testing can cover many more combinations of factors.

Acceptance testing: test basis

How do we know that the system is ready to be used for real? Examples of work products that can be used as a test basis for the various forms of acceptance testing include:

- business processes
- user or business requirements
- regulations, legal contracts and standards
- use cases
- system requirements
- system or user documentation
- installation procedures
- risk analysis reports.

For operational acceptance testing (OAT), there are some additional aspects with specific work products that can be a test basis:

- backup and restore/recovery procedures
- disaster recovery procedures

- non-functional requirements
- operations documentation
- deployment and installation instructions
- performance targets
- database packages
- security standards or regulations.

Note that it is particularly important to have very clear, well-tested and frequently rehearsed procedures for disaster recovery and restoring backups. If you are in the situation of having to perform these procedures, then you may be in a state of panic, since something serious will have already gone wrong. In that psychological state, it is very easy to make mistakes, and here mistakes could be disastrous. There are stories of organizations who compounded one disaster by accidentally deleting their backups, or who find that their backups are unusable or incomplete! This is why restoring from backups is an important test to do regularly.

Acceptance testing: test objects

What are we actually testing at this level? The emphasis here is in gaining confidence, based on the particular form of acceptance testing: user confidence, confidence in operations, confidence that we have met legal or regulatory requirements, and confidence that real users will like and be happy with the software we are selling. Some of the things we are testing are similar to the test objects of system testing. Typical test objects for acceptance testing include:

- system under test (SUT)
- system configuration and configuration data
- business processes for a fully integrated system
- recovery systems and hot sites (for business continuity and disaster recovery testing)
- operational and maintenance processes
- forms
- reports
- existing and converted production data.

Acceptance testing: typical defects and failures

Examples of defects and failures that can typically be revealed by acceptance testing include:

- System workflows do not meet business or user requirements.
- Business rules are not implemented correctly.
- System does not satisfy contractual or regulatory requirements.
- Non-functional failures such as security vulnerabilities, inadequate performance efficiency under high load, or improper operation on a supported platform.

At the end of the description of all the test levels, refer to Table 2.1, which summarizes the characteristics of each test level.

Acceptance testing: specific approaches and responsibilities

The acceptance test should answer questions such as: 'Can the system be released?', 'What, if any, are the outstanding (business) risks?' and 'Has development met their obligations?' Acceptance testing is most often the responsibility of the user or customer, although other stakeholders may be involved as well. The execution of the acceptance test requires a test environment that is, for most aspects, representative of the production environment ('as-if production').

The goal of acceptance testing is to establish confidence in the system, part of the system or specific non-functional characteristics, for example usability of the system. Acceptance testing is most often focused on a validation type of testing, where we are trying to determine whether the system is fit for purpose. Finding defects should not be the main focus in acceptance testing. Although it assesses the system's readiness for deployment and use, it is not necessarily the final level of testing. For example, a large-scale system integration test may come after the acceptance of a system.

Acceptance testing may occur at more than just a single level, for example:

- A COTS software product may be acceptance tested when it is installed or integrated.
- Acceptance testing of the usability of a component may be done during component testing.
- Acceptance testing of a new functional enhancement may come before system testing.

User acceptance testing focuses mainly on the functionality, thereby validating the fitness for use of the system by the business user, while the operational acceptance test (also called production acceptance test) validates whether the system meets the requirements for operation. The user acceptance test is performed by the users and application managers. In terms of planning, the user acceptance test usually links tightly to the system test, and will, in many cases, be organized partly overlapping in time. If the system to be tested consists of a number of more or less independent subsystems, the acceptance test for a subsystem that meets its exit criteria from the system test can start while another subsystem may still be in the system test phase. In most organizations, system administration will perform the operational acceptance test shortly before the system is released. The operational acceptance test may include testing of backup/restore, data load and migration tasks, disaster recovery, user management, maintenance tasks and periodic check of security vulnerabilities.

Note that organizations may use other terms, such as factory acceptance testing and site acceptance testing for systems that are tested before and after being moved to a customer's site.

In iterative development, different forms of acceptance testing may be done at various times, and often in parallel. At the end of an iteration, a new feature may be tested to validate that it meets stakeholder and user needs. This is user acceptance testing. If software for general release (COTS) is being developed, alpha and beta testing may be used at or near the end of an iteration or set of iterations. Operational and regulatory acceptance testing may also occur at the end of an iteration or set of iterations.

Test level characteristics: summary

Table 2.1 summarizes the characteristics of the different test levels: the test basis, test objects and typical defects and failures. We have covered these in the various sections, but it is useful to contrast them in order to distinguish them from each other. We have omitted some of the detail to make the table easier to take in at a glance. Note that some of the typical defects for integration testing are only for system integration testing (SIT).

TABLE 2.1 Test level characteristics

	Component testing	Integration testing	System testing	Acceptance testing
Objectives	• reduce risk • verify functional and non-functional behaviour • build confidence in components • find defects • prevent defects to higher levels	• reduce risk • verify functional and non-functional behaviour • build confidence in interfaces • find defects • prevent defects to higher levels	• reduce risk • verify functional and non-functional behaviour • validate completeness, works as expected • build confidence in whole system • find defects • prevent defects to higher levels	• establish confidence in whole system and its use • validate completeness, works as expected • verify functional and non-functional behaviour
Test basis	• detailed design • code • data models • component specifications	• software/system design • sequence diagrams • interface and communication protocol specs • use cases • architecture (component or system) • workflows • external interface definitions	• requirement specs (functional and non-functional) • risk analysis reports • use cases • epics and user stories • models of system behaviour • state diagrams • system and user manuals	• business processes • user, business, system requirements • regulations, legal contracts and standards • use cases • documentation • installation procedures • risk analysis
Test objects	• components, units, modules • code • data structures • classes • database models	• subsystems • databases • infrastructure • interfaces • APIs • microservices	• applications • hardware/software • operating systems • system under test • system configuration and data	• system under test (SUT) • system configuration and data • business processes

(Continued)

TABLE 2.1   Test level characteristics (*Continued*)

	Component testing	Integration testing	System testing	Acceptance testing
Test objects				• recovery systems • operation and maintenance processes • forms • reports • existing and converted production data
Typical defects and failures	• wrong functionality • data flow problems • incorrect code/ logic	• data problems • inconsistent message structure (SIT) • timing problems • interface mismatch • communication failures • incorrect assumptions • not complying with regulations (SIT)	• incorrect calculations • incorrect or unexpected behaviour • incorrect data/ control flows • cannot complete end-to-end tasks • does not work in production environment(s) • not as described in manuals/ documentation	• system workflows do not meet business or user needs • business rules not correct • contractual or regulatory problems • non-functional failures (performance, security)

2.2.2 Test types

In this section, we will look at different test types. We will discuss tests that focus on the functionality of a system, which informally is *testing what the system does*. We will also discuss tests that focus on non-functional attributes of a system, which informally is *testing how well the system does what it does*. We will introduce testing based on the system's structure. Finally, we will look at testing of changes to the system, both confirmation testing (testing that the changes succeeded) and regression testing (testing that the changes did not affect anything unintentionally).

The test types discussed here can involve the development and use of a model of the software or its behaviours. Such models can occur in structural testing when we use control flow models or menu structure models. Such models in non-functional testing can involve performance models, usability models and security threat models.

They can also arise in functional testing, such as the use of process flow models, state transition models or plain language specifications. Examples of such models will be found in Chapter 4.

As we go through this section, watch for the Syllabus terms **black-box testing**, **confirmation testing**, **functional testing**, **non-functional testing**, **regression testing**, **test type** and **white-box testing**. You will find these terms defined in the Glossary as well.

Test types are introduced as a means of clearly defining the objective of a certain test level for a program or project. We need to think about different types of testing because testing the functionality of the component or system may not be sufficient at each level to meet the overall test objectives. Focusing the testing on a specific test objective and, therefore, selecting the appropriate type of test, helps make it easier to make and communicate decisions about test objectives. Typical objectives may include:

- Evaluating functional quality, for example whether a function or feature is complete, correct and appropriate.

- Evaluating non-functional quality characteristics, for example reliability, performance efficiency, security, compatibility and usability.

- Evaluating whether the structure or architecture of the component or system is correct, complete and as specified.

- Evaluating the effects of changes, looking at both the changes themselves (for example, defect fixes) and also the remaining system to check for any unintended side-effects of the change. These are confirmation testing and regression testing, respectively, and are discussed in Section 2.2.3.

A **test type** is focused on a particular test objective, which could be the testing of a function to be performed by the component or system; a non-functional quality characteristic, such as reliability or usability; the structure or architecture of the component or system; or related to changes, that is, confirming that defects have been fixed (confirmation testing, or re-testing) and looking for unintended changes (regression testing). Depending on its objectives, testing will be organized differently. For example, component testing aimed at performance would be quite different from component testing aimed at achieving decision coverage.

> **Test type** A group of test activities based on specific test objectives aimed at specific characteristics of a component or system.

2.2.2.1 *Functional testing*

The function of a system (or component) is what it does. This is typically described in work products such as business requirements specifications, functional specifications, use cases, epics or user stories. There may be some functions that are assumed to be provided that are not documented. They are also part of the requirements for a system, though it is difficult to test against undocumented and implicit requirements. Functional tests are based on these functions, described in documents or understood by the testers, and may be performed at all test levels (for example, tests for components may be based on a component specification).

Functional testing considers the specified behaviour and is often also referred to as black-box testing (specification-based testing). This is not entirely true, since black-box testing also includes non-functional testing (refer to Section 2.3.2).

Functional testing can also be done focusing on suitability, interoperability testing, security, accuracy and compliance. Security testing, for example, investigates the functions (for example, a firewall) relating to detection of threats, such as viruses, from malicious outsiders.

> **Functional testing** Testing performed to evaluate if a component or system satisfies functional requirements.

Testing of functionality could be done from different perspectives, the two main ones being requirements-based or business-process-based.

Requirements-based testing uses a specification of the functional requirements for the system as the basis for designing tests. A good way to start is to use the table of contents of the requirements specification as an initial test inventory or list of items to test (or not to test). We should also prioritize the requirements based on risk criteria (if this is not already done in the specification) and use this to prioritize the tests. This will ensure that the most important and most critical tests are included in the testing effort.

Business-process-based testing uses knowledge of the business processes. Business processes describe the scenarios involved in the day-to-day business use of the system. For example, a personnel and payroll system may have a business process along the lines of:

- Someone joins the company.
- They are paid on a regular basis.
- They finally leave the company.

User scenarios originate from object-oriented development but are nowadays popular in many development life cycles. They also take the business processes as a starting point, although they start from tasks to be performed by users. Use cases are a very useful basis for test cases from a business perspective.

The techniques used for functional testing are often specification-based, but experience-based techniques can also be used (refer to Chapter 4 for more on test techniques). Test conditions and test cases are derived from the functionality of the component or system. As part of test design, a model may be developed, such as a process model, state transition model or a plain-language specification.

The thoroughness of functional testing can be measured by a coverage measure based on elements of the function that we can list. For example, we can list all of the options available from every pull-down menu. If our set of tests has at least one test for each option, then we have 100% coverage of these menu options. Of course, that does not mean that the system or component is 100% tested, but it does mean that we have at least touched every one of the things we identified. When we have traceability between our tests and functional requirements, we can identify which requirements we have not yet covered, that is, have not yet tested (coverage gaps). For example, if we covered only 90% of the menu options, we could add tests so that the untested 10% are then covered.

Special skills or knowledge may be needed for functional testing, particularly for specialized application domains. For example, medical device software may need medical knowledge both for the design and testing of such systems. The worst thing that a heart pacemaker can do is not to stop giving the electrical stimulant to the heart (the heart may still limp along less efficiently). The worst thing is to speed up, giving the signal much too frequently; this can be fatal. Other specialized application areas include gaming or interactive entertainment systems, geological modelling for oil and gas exploration, or automotive systems.

2.2.2.2 Non-functional testing

This test type is the testing of the quality characteristics, or non-functional attributes of the system (or component or integration group). Here we are interested in how well or how fast something is done. We are testing something that we need to measure on a scale of measurement, for example time to respond.

Non-functional testing, as functional testing, is performed at all test levels. Non-functional testing includes, but is not limited to, performance testing, load testing, stress testing, usability testing, maintainability testing, reliability testing, portability testing and security testing. It is the testing of how well the system works.

Many have tried to capture software quality in a collection of characteristics and related sub-characteristics. In these models, some elementary characteristics keep on reappearing, although their place in the hierarchy can differ. The International Organization for Standardization (ISO) has defined a set of quality characteristics in ISE/IEC 25010 [2011]. The ISO/IEC 25010 standard provides the following classification of the non-functional software quality characteristics:

Non-functional testing Testing performed to evaluate that a component or system complies with non-functional requirements.

- performance efficiency
- compatibility
- usability
- reliability
- security
- maintainability
- portability.

A common misconception is that non-functional testing occurs only during higher levels of testing such as in a system test, system integration test and acceptance test. In fact, non-functional testing may be performed at all test levels; the higher the level of risk associated with each type of non-functional testing, the earlier in the life cycle it should occur. Ideally, non-functional testing involves tests that quantifiably measure characteristics of the systems and software. For example, in performance testing we can measure transaction throughput, resource utilization and response times. Generally, non-functional testing defines expected results in terms of the external behaviour of the software. This means that we typically use black-box test techniques. For example, we could use boundary value analysis to define the stress conditions for performance tests, and equivalence partitioning to identify types of devices for compatibility testing, or to identify user groups for usability testing (novice, experienced, age range, geographical location, educational background).

The thoroughness of non-functional testing can be measured by the coverage of non-functional elements. If we had at least one test for each major group of users, then we would have 100% coverage of those user groups that we had identified. Of course, we may have forgotten an important user group, such as those with disabilities, so we have only covered the groups we have identified.

If we have traceability between non-functional tests and non-functional requirements, we may be able to identify coverage gaps. For example, an implicit requirement is for accessibility for disabled users.

Special skills or knowledge may be needed for non-functional testing, such as for performance testing, usability testing or security testing (for example, for specific development languages).

More about non-functional testing is found in other ISTQB qualification syllabi, including the Advanced Test Analyst, the Advanced Technical Test Analyst, and the specialist syllabi Security Tester, Performance Testing, and Usability Testing.

2.2.2.3   Black-box testing

Black-box testing (refer to Section 4.2), also known as specification-based, is a software testing approach that focuses on testing a software application without requiring knowledge of the internal code or implementation details. Testers approach the system as a 'black box', meaning they are only concerned with inputs, outputs and the system's behaviour, without any understanding of the internal logic, structure or code. **Black-box** testing can be applied at various levels of software development, including system testing, acceptance testing and integration testing. Different techniques can be employed in black-box testing, such as equivalence partitioning, boundary value analysis and state transition testing. These techniques help create test cases that are representative of different input scenarios. To some extent black-box testing promotes independence between the testing team and the development team since testers do not need knowledge of the internal codebase, allowing for a separation of concerns between testing and development activities.

> **Black-box testing** Testing based on an analysis of the specification of the component or system (also known as specification-based testing).

2.2.2.4   White-box testing

The fourth test type looks at the internal structure or implementation of the system or component. If we are talking about the structure of a system, we may call it the system architecture. Structural elements also include the code itself, control flows, business processes and data flows. **White-box testing** (refer to Section 4.3) is also referred to as structural testing or glass-box because we are interested in what is happening inside the box.

White-box testing is most often used as a way of measuring the thoroughness of testing through the coverage of a set of structural elements or coverage items. It can occur at any test level, although it is true to say that it tends to be mostly applied at component testing and component integration testing, and generally is less likely at higher test levels, except for business process testing. At component integration level it may be based on the architecture of the system, such as a calling hierarchy or the interfaces between components (the interfaces themselves can be listed as coverage items). The test basis for system, system integration or acceptance testing could be a business model, for example business rules.

> **White-box testing** Testing based on an analysis of the internal structure of the component or system (also known as clear-box testing, code-based testing, glass-box testing, logic-coverage testing, logic-driven testing, structural testing, structure-based testing).

At component level, and to a lesser extent at component integration testing, there is good tool support to measure code coverage. Coverage measurement tools assess the percentage of executable elements (for example, statements or decision outcomes) that have been exercised (that is, they have been covered) by a test suite. If coverage is not 100%, then additional tests may need to be written and run to cover those parts that have not yet been exercised. This of course depends on the exit criteria. (Coverage and white-box test techniques are covered in Chapter 4.)

Special skills or knowledge may be needed for white-box testing, such as knowledge of the code (to interpret coverage tool results) or how data is stored (for database queries).

2.2.3 Confirmation testing and regression testing

The final test type is the testing of changes. This category is slightly different from the others because if you have made a change to the software, you will have changed the way it functions, how well it functions (or both) and its structure. However, we are looking here at the specific types of tests relating to changes, even though they may include all of the other test types. There are two things to be particularly aware of when changes are made: the change itself and any other effects of the change.

Confirmation testing (re-testing)

When a test fails and we determine that the cause of the failure is a software defect, the defect is reported, and we can expect a new version of the software that has had the defect fixed. In this case we will need to execute the test again to confirm that the defect has indeed been fixed. This is known as **confirmation testing** (also known as re-testing).

When doing confirmation testing, it is important to ensure that steps leading up to the failure are carried out in exactly the same way as described in the defect report, using the same inputs, data and environment, and possibly extending beyond the test to ensure that the change has indeed fixed all of the problems due to the defect. If the test now passes, does this mean that the software is now correct? Well, we now know that at least one part of the software is correct—where the defect was. But this is not enough. The fix may have introduced or uncovered a different defect elsewhere in the software. The way to detect these unexpected side-effects of fixes is to do regression testing.

> **Confirmation testing** A type of change-related testing performed after fixing a defect to confirm that a failure caused by that defect does not re-occur. (also known as re-testing).

Regression testing

Like confirmation testing, **regression testing** involves executing test cases that have been executed before. The difference is that, for regression testing, the test cases probably passed the last time they were executed (compare this with the test cases executed in confirmation testing—they failed the last time).

The term regression testing is something of a misnomer. It would be better if it were called anti-regression testing because we are executing tests with the intent of checking that the system has *not* regressed (that is, it does not now have more defects in it as a result of some change). More specifically, the purpose of regression testing is to make sure (as far as is practical) that modifications in the software or the environment have not caused unintended adverse side effects and that the system still meets its requirements.

> **Regression testing** A type of change-related testing to detect whether defects have been introduced or uncovered in unchanged areas of the software.

It is common for organizations to have what is usually called a regression test suite or regression test pack. This is a set of test cases that is specifically used for regression testing. They are designed to collectively exercise most functions (certainly the most important ones) in a system, but not test any one in detail. It is appropriate to have a regression test suite at every level of testing (component testing, integration testing, system testing, etc.). In some cases, all of the test cases in a regression test suite would be executed every time a new version of software is produced; this makes them ideal candidates for automation. However, it is much better to be able to select subsets for execution, especially if the regression test suite is very large. In Agile development, a selection of regression tests would be run to meet the objectives of a particular iteration. Automation of regression tests should start as early as possible in the project. Refer to Chapter 6 for more on test automation.

Regression tests are executed whenever the software changes, either as a result of fixes or new or changed functionality. It is also a good idea to execute them when some aspect of the environment changes, for example when a new version of the host operating system is introduced, or the production environment has a new version of the Java Virtual Machine or anti-malware software.

Maintenance of a regression test suite should be carried out, so it evolves over time in line with the software. As new functionality is added to a system, new regression tests should be added. As old functionality is changed or removed, so too should regression tests be changed or removed. As new tests are added, a regression test suite may become very large. If all the tests have to be executed manually it may not be possible to execute them all every time the regression suite is used. In this case, a subset of the test cases has to be chosen. This selection should be made considering the latest changes that have been made to the software. Sometimes a regression test suite of automated tests can become so large that it is not always possible to execute them all. It may be possible and desirable to eliminate some test cases from a large regression test suite, for example if they are repetitive (tests which exercise the same conditions) or can be combined (if they are always run together). Another approach is to eliminate test cases when the risk associated with that test is so low that it is not worth running it anymore.

Both confirmation testing and regression testing are done at all test levels.

In iterative and incremental development, changes are more frequent, even continuous, and the software is refactored frequently. This makes confirmation testing and regression testing even more important. But iterative development such as Agile should also include continuous testing, and this testing is mainly regression testing. For internet of things (IoT) systems, change-related testing covers not only software systems but also the changes made to individual objects or devices, which may be frequently updated or replaced.

2.3 MAINTENANCE TESTING

> **SYLLABUS LEARNING OBJECTIVES FOR 2.3 MAINTENANCE TESTING (K2)**
>
> **FL-2.3.1**   **Summarize maintenance testing and its triggers (K2)**

Maintenance testing Testing the changes to an operational system or the impact of a changed environment to an operational system.

Once deployed, a system is often in service for years or even decades. During this time, the system and its operational environment are often corrected, changed or extended. As we go through this section, watch for the Syllabus term **maintenance testing**. You will find this term also defined in the Glossary and the ISTQB online Glossary.

Testing that is executed during this life cycle phase is called **maintenance testing**. Maintenance testing, along with the entire process of maintenance releases, should be carefully planned. Not only must planned maintenance releases be considered, but the process for developing and testing hot fixes must be as well. Maintenance testing includes any type of testing of changes to an existing, operational system, whether the changes result from modifications, migration or retirement of the software or system.

Modifications can result from planned enhancement changes, such as those referred to as minor releases, that include new features and accumulated (non-emergency) bug fixes. Modifications can also result from corrective and more urgent emergency changes. Modifications can also involve changes of environment, such as planned operating system or database upgrades, planned upgrade of COTS software, or patches to correct newly exposed or discovered vulnerabilities of the operating system.

Migration involves moving from one platform to another. This can involve abandoning a platform no longer supported or adding a new supported platform.

Either way, testing must include operational tests of the new environment as well as of the changed software. Migration testing can also include conversion testing, where data from another application will be migrated into the system being maintained.

Note that maintenance testing is different from testing for maintainability (which is the degree to which a component or system can be modified by the intended maintainers). In this section, we will discuss maintenance testing.

The same test process steps will apply as for testing during development and, depending on the size and risk of the changes made, several levels of testing are carried out: a component test, an integration test, a system test and an acceptance test. If testing is done more formally, an application for a change may be used to produce a test plan for testing the change, with test cases changed or created as needed. In less formal testing, thought needs to be given to how the change should be tested, even if this planning, updating of test cases and execution of the tests is part of a continuous process.

The scope of maintenance testing depends on several factors, which influence the test types and test levels. The factors are:

- Degree of risk of the change, for example a self-contained change is a lower risk than a change to a part of the system that communicates with other systems.

- The size of the existing system, for example a small system would need less regression testing than a larger system.

- The size of the change, which affects the amount of testing of the changes that would be needed. The amount of regression testing is more related to the size of the system than the size of the change.

2.3.1 Maintenance testing and its triggers

Triggers for maintenance

As stated, maintenance testing is done on an existing operational system. There are three possible triggers for maintenance testing:

- modifications
- migration or upgrades
- retirement.

Modifications include planned enhancement changes (for example, release-based), corrective and emergency changes and changes of environment, such as planned operating system or database upgrades, or patches to newly exposed or discovered vulnerabilities of the operating system and upgrades of COTS software.

Modifications may also be of hardware or devices, not just software components or systems. For example, in IoT systems, new or significantly modified hardware devices may be introduced to a working system. The emphasis in maintenance testing would likely focus on different types of integration testing and security testing at all test levels.

Maintenance testing for migration (for example, from one platform to another) should also include operational testing of the new environment, as well as the changed software. It is important to know that the platform you will be transferring to is sound before you start migrating your own files and applications.

Maintenance testing for the retirement of a system may include the testing of data migration or archiving if long data-retention periods are required. Testing of restore or retrieve procedures after archiving may also be needed. There is no point in trying to save and preserve something that you can no longer access. These procedures should be regularly tested and action taken to migrate away from technology that is reaching the end of its life. You may remember seeing magnetic tape on old movies, which was thought to be a good long-term archiving solution at the time.

Impact analysis and regression testing

As mentioned earlier, maintenance testing usually consists of two parts:

- Testing the changes.
- Regression tests to show that the rest of the system has not been affected by the maintenance work.

In addition to testing what has been changed, maintenance testing includes extensive regression testing to parts of the system that have not been changed. Some systems will have extensive regression suites (automated or not) where the costs of executing all of the tests would be significant. A major and important activity within maintenance testing is impact analysis. During impact analysis, together with stakeholders, a decision is made on what parts of the system may be unintentionally affected and therefore need more extensive regression testing. Risk analysis will help to decide where to focus regression testing. It is unlikely that the team will have time to repeat all the existing tests, so this gives us the best value for the time and effort we can spend in regression testing.

If the test specifications from the original development of the system are kept, you may be able to reuse them for regression testing and to adapt them for changes to the system. This may be as simple as changing the expected results for your existing tests. Sometimes additional tests may need to be built. Extension or enhancement to the system may mean new areas have been specified and tests would be drawn up just as for the development. Do not forget that automated regression tests will also need to be updated in line with the changes; this can take significant effort, depending on the architecture of your automation.

Impact analysis can also be used to help make a decision about whether or not a particular change should be made. If the change has the potential to cause high-risk vulnerabilities throughout the system, it may be a better decision not to make that change.

There are a number of factors that make impact analysis more difficult:

- Specifications are out of date or missing (for example, business requirements, user stories, architecture diagrams).
- Test cases are not documented or are out of date.
- Bi-directional traceability between tests and the test basis has not been maintained.
- Tool support is weak or non-existent.
- The people involved do not have domain and/or system knowledge.
- The maintainability of the software has not been taken into enough consideration during development.

Impact analysis can be very useful in making maintenance testing more efficient, but if it is not, or cannot be, done well, then the risks of making the change are greatly increased.

CHAPTER REVIEW

Let us review what you have learned in this chapter.

From Section 2.1, you should now understand the relationship between development activities and test activities within a development life cycle and be familiar with sequential life cycle models (waterfall and V-model) and iterative/incremental life cycle models (Unified Process, Scrum, Kanban and Spiral). You should understand the impact of the choice for a software development lifecycle on testing and recall characteristics of good testing in any life cycle model. You should be able to recall the examples of test-first approaches to development and summarize how DevOps has an impact on software testing. Finally, you should be able to explain the shift-left approach and how retrospectives can be used as a mechanism for (test) process improvement. You should know the Glossary terms **shift-left** and **test level**.

From Section 2.2, you should know the typical levels of testing component, integration (both component integration and system integration), system and acceptance testing. You should be able to compare the different levels of testing with respect to their major objectives, the test basis, typical objects of testing, typical defects and failures, and approaches and responsibilities for each test level. You should know the Glossary terms **acceptance testing**, **component integration testing**, **component testing**, **integration testing**, **system integration testing**, **system testing**, **and test object**.

Also from Section 2.2, you should know the five major types of test (functional, non-functional, black-box, white-box and change-related) and should be able to provide some concrete examples for each of these. You should understand that functional and white-box tests occur at any test level and be able to explain how they are applied in the various test levels. You should be able to identify and describe non-functional test types based on non-functional requirements and product quality characteristics. Finally, you should be able to explain the purpose of confirmation testing (re-testing) and regression testing in the context of change-related testing. You should know the Glossary terms **black-box testing**, **confirmation testing**, **functional testing**, **non-functional testing**, **regression testing**, **test type** and **white-box testing**.

From Section 2.3, you should be able to compare maintenance testing to testing of new applications. You should be able to identify triggers and reasons for maintenance testing, such as modifications, migration and retirement. You should know the Glossary term **maintenance testing**.

SAMPLE EXAM QUESTIONS

Question 1 Which of the following statements is true?

a. Overlapping test levels and test activities are more common in sequential life cycle models than in iterative incremental models.

b. The V-model is an iterative incremental life cycle model because each development activity has a corresponding test activity.

c. When completed, iterative incremental life cycle models are more likely to deliver the full set of features originally envisioned by stakeholders than sequential models.

d. In iterative and incremental life cycle models, delivery of usable software to end-users is much more frequent than in sequential models.

Question 2 In which test-first approach is the desired behaviour of an application expressed in test cases, whereby the test cases are usually written using the Given/When/Then format?

a. Test-driven development.

b. Acceptance test-driven development.

c. Behaviour driven-development.

d. Feature driven-development.

Question 3 Which of the following is a test type?

a. Component testing.

b. Functional testing.

c. System testing.

d. Acceptance testing.

Question 4 Consider the three triggers for maintenance, and match the event with the correct trigger:

1. Data conversion from one system to another.

2. Upgrade of COTS software.

3. Test of data archiving.

4. System now runs on a different platform and operating system.

5. Testing restore or retrieve procedures.

6. Patches for security vulnerabilities.

a. Modification: 2 and 3, Migration: 1 and 5, Retirement: 4 and 6.

b. Modification: 2 and 6, Migration: 1 and 4, Retirement: 3 and 5.

c. Modification: 2 and 4, Migration: 1 and 3, Retirement: 5 and 6.

d. Modification: 1 and 5, Migration: 2 and 3, Retirement: 4 and 6.

Question 5 Which of these is a functional test?

a. Measuring response time on an online booking system.

b. Checking the effect of high volumes of traffic in a call centre system.

c. Checking the online bookings screen information and the database contents against the information on the letter to the customers.

d. Checking how easy the system is to use, particularly for users with disabilities such as impaired vision.

Question 6 Which of the following is true, regarding the process of testing emergency fixes?

a. There is no time to test the change before it goes live, so only the best developers should do this work and should not involve testers as they slow down the process.

b. Just run the retest of the defect actually fixed.

c. Always run a full regression test of the whole system in case other parts of the system have been adversely affected.

d. Retest the changed area and then use risk assessment to decide on a reasonable subset of the whole regression test to run in case other parts of the system have been adversely affected.

Question 7 A regression test:

a. Is only run once.

b. Will always be automated.

c. Will check unchanged areas of the software to see if they have been affected.

d. Will check changed areas of the software to see if they have been affected.

Question 8 Non-functional testing includes:

a. Testing to see where the system does not function correctly.

b. Testing the quality attributes of the system including reliability and usability.

c. Gaining user approval for the system.

d. Testing a system feature using only the software required for that function.

Question 9 Which of the following is an example of a practice that illustrates a shift-left approach in testing?

a. Performing retrospectives to improve the quality of the test basis.

b. Defining specific and different test objectives for each test level.

c. Performing static analysis on source code prior to dynamic testing.

d. Minimizing the regression risk due to a large scale of automated regression tests.

CHAPTER THREE
Static testing

Static test techniques provide a powerful way to improve the quality and productivity of software development. This chapter describes static test techniques, including reviews, and provides an overview of how they are conducted. The fundamental objective of static testing is to improve the quality of software work products by assisting engineers to recognize and fix their own defects early in the software development process. While static testing techniques will not solve all the problems, they are enormously effective. Static techniques can improve both quality and productivity by impressive factors. Static testing is not magic, and it should not be considered a replacement for dynamic testing, but all software organizations should consider using reviews in all major aspects of their work, including requirements, design, implementation, testing and maintenance. Static analysis tools implement automated checks, for example on code.

3.1 STATIC TESTING BASICS

> **SYLLABUS LEARNING OBJECTIVES FOR 3.1 STATIC TESTING BASICS (K2)**
>
> **FL-3.1.1** Recognize types of products that can be examined by the different static test techniques (K1)
>
> **FL-3.1.2** Explain the value of static testing (K2)
>
> **FL-3.1.3** Compare and contrast static and dynamic testing (K2)

In this section, we consider how static testing techniques fit into the overall test process. Dynamic testing requires that we run the item or system under test, but static testing techniques allow us to find defects directly in work products, without the execution of the code and without the need to isolate the failure to locate the underlying defect. Static techniques include both reviews and static analysis, each of which we'll discuss in this chapter. Static techniques are efficient ways to find and remove defects and can find certain defects that are hard to find with dynamic testing. As we go through this section, watch for the Syllabus terms **dynamic testing**, **static analysis** and **static testing**. You will find these terms also defined in the Glossary.

In Chapter 1, we saw that testing is defined as all life cycle activities, both static and dynamic, to do with planning, preparing and evaluating software and related work products. As indicated in that definition, two approaches can be used to achieve these objectives: static testing and dynamic testing. Static analysis is a form of automated static testing.

The definitions of **static analysis** and **static testing** are very similar and, to be honest, are somewhat confusing! The definition of static analysis would apply equally well to reviews, which are a form of static testing but are not part of static analysis. Generally, static analysis involves the use of tools to do the analysis; in fact, the Syllabus describes reviews as a manual process and static analysis being with the support of a tool. However, reviews can also be supported by a tool (for the review process); static analysis is <u>performed</u> by a tool, not just supported by one. The key difference is: considering the code we want to evaluate, **dynamic testing** actually executes that code; static testing (including static analysis) does NOT execute the code we are evaluating.

The objectives for both types of static testing include not just finding defects (assessing correctness) but also assessing characteristics such as readability, completeness, testability and consistency. Static techniques can be used for both verification and validation.

In Agile development, static testing is used when developers and business representatives work together on the system, particularly in planning what the system should do. During example mappings, collaborative user story writing and backlog refinement, using reviews can help to ensure that the user stories and other work products meet the criteria defined for them, such as the definition of ready (refer to Section 5.1.3). But even more importantly, reviews can help to ensure that the user stories are the right user stories from a user perspective (validation). Reviews of user stories can focus on completeness, understandability and testability, as well as on evaluating acceptance criteria. One of the key roles for the tester in the team is to ask the right questions, to explore alternatives (what could go wrong) and to challenge assumptions.

Static analysis is often part of the Continuous Integration (CI) framework of the delivery pipeline (refer to Section 2.1.4). Although static analysis is more often used to detect code defects (and standards violations), it can also help to evaluate maintainability, reliability and security. Even tools such as grammar checkers and spell-checkers are a form of static analysis for written documents.

One area where static analysis is often used (and is critical for) is in safety-critical systems such as flight control software, medical devices or nuclear power control software. However, static analysis is also very important in security testing, as it can identify malicious code (which does not make itself visible in execution). In continuous delivery and continuous deployment, the automated build systems also frequently make use of static analysis as part of the build process.

Because static testing can be used earlier than dynamic testing (where you need something to run the tests on), problems can be identified earlier in the life cycle and in the iteration. It may take less effort to fix the problems found because the user story, code or work product will be fresher in the mind of the developer.

Static analysis The process of evaluating a component or system without executing it, based on its form, structure, content or documentation.

Static testing Testing that does not involve the execution of a test item.

Dynamic testing Testing that involves the execution of the test item.

3.1.1 Work products examinable by static testing

Static analysis is most often used to evaluate code against various criteria, such as adherence to coding standards, thresholds for complexity, spelling and grammar correctness and reading difficulty (the last few for work products other than code). However, static testing is broader than automated evaluations; reviews can be applied to any type of work product, including:

- Any type of specification: business requirements, functional requirements, security requirements.
- Epics, user stories and acceptance criteria.
- Source code.
- Testware, that is, any type of work product to do with testing, for example test plans, test conditions, test charters, test cases, test procedures and automated test scripts.
- Product backlog items.
- User guides, help text, wizards and other things designed to help the user to more effectively use the system, and any other project documentation.
- Web pages (there are also static analysis tools to analyze whether any links are broken, for example).
- Contracts (a particularly important work product to review, as a lot of money may be riding on the specific wording), project plans, schedules and budgets.
- Models such as activity diagrams or other models used in model-based testing (MBT).

Reviews apply to any work product; the reviewers need to be able to read and understand it but can then provide feedback about it. Refer to Section 3.2 for more on reviews.

Static analysis of software code is done using a tool to analyze the code with respect to the criteria of interest. For example, static analysis tools can identify dead code, a section of code that can never be reached from anywhere else in the code. This dead code can never be executed and so should be removed, as it could be confusing to leave it in. Static analysis can also identify a variable whose value is used before it has been defined. This type of defect can cause failures which are difficult to find by dynamic testing.

Static analysis needs to work against some kind of structure. Source code is checked against various rules for code elements such as statements, branches and other structures. These rules may be configurable in the static analysis tool in terms of the application and the severity of different rule violations. There are also tools that can analyze models, text with a formal syntax, or even natural language text, for example in requirements, to catch some typos, assess readability level (ease of understandability), etc., so these are also a form of static analysis.

Reviews are generally not appropriate for work products that are difficult to interpret by human beings, such as object code; object code from a third-party supplier can typically not be analyzed using static analysis tools for legal reasons. There may be exceptions to this, for example using tools on object code to check for patent infringement in a lawsuit, by reverse engineering the code.

3.1.2 Value of static testing

Studies have shown that as a result of reviews, a significant increase in productivity and product quality can be achieved; refer to Gilb and Graham [1993] and van Veenendaal [1999]. Reducing the number of defects early in the product life cycle also means that less time would be spent on testing and maintenance. The use of static testing, such as reviews on software work products, has a number of advantages:

- Since static testing can start early in the life cycle, early feedback on quality issues can be established, for example an early validation of user requirements rather than late in the life cycle during acceptance testing. Feedback during design review or backlog refinement is more useful than after a feature has been built. This illustrates how the principle of early testing (Section 1.3) is applied.

- By detecting defects at an early stage, rework costs are most often relatively low, and thus relatively cheap improvements to the quality of software products can be achieved, as many of the follow-on costs of late updates are avoided, for example additional regression tests, confirmation tests, etc.

- Defects are more efficiently detected and corrected, particularly since this is done before dynamic test execution.

- Defects that are not easily found by dynamic testing can be identified, such as unreachable ('dead') code, designs not implemented as desired, defects in non-executable work products such as user guides, and security vulnerabilities.

- Defects in future design and code can be prevented by uncovering inconsistencies, ambiguities, contradictions, omissions, inaccuracies and redundancies in requirements or user stories.

- Since rework effort is substantially reduced, development productivity figures are likely to increase.

- Reduced development cost and time due to reduced rework later on. Defects found by reviews, or a static analysis tool can be fixed immediately at the lowest cost and in the quickest time.

- Reduced testing cost and time. If defects are found and fixed before test execution starts, there are fewer to find in testing, so more tests pass, and there are fewer defect reports to write and fewer confirmation tests to run after fixes. This saves both time and money.

- Reduced total cost of quality over the software's lifetime. If defects are found and fixed early, then there should be fewer that get through to later testing or operation. The defects that are not there do not need to be investigated or fixed, saving time and money. There is a perception that reviews are costly to implement, but the question to ask is, 'Compared to what?' Good static testing saves money.

- Improved communication within the team, since there is an exchange of information between the participants during reviews, which can lead to an increased awareness of quality issues. Stakeholders can ensure that requirements meet their actual needs, and a shared understanding can be created among stakeholders. It is recommended to involve a wide variety of stakeholders in static testing, mainly reviews.

In conclusion, static testing is a very suitable method for improving the quality of software work products. This applies primarily to the assessed work products themselves. It is also important that the quality improvement is not achieved just once but has a more permanent nature. The feedback from the static testing process to the development process allows for process improvement, which supports the avoidance of similar errors being made in the future. This is particularly useful for sprint or project retrospectives.

3.1.3 Differences between static and dynamic testing

Static and dynamic testing have the same objectives: to assess the quality of work products and identify defects as early as possible. But static and dynamic testing are not the same. They find different types of defect, so they are complementary and are best used together. It does not make sense to ask if one is better than the other; both are useful and needed.

With dynamic testing methods, software is executed using a set of input values and its output is then examined and compared to what is expected. During static testing, software work products are examined manually, or with a set of tools, but not executed. Dynamic testing can be started early by identifying test conditions and test cases as early as possible in the life cycle (as we discussed in Chapter 1 Section 1.4), but dynamic test execution can only be applied to software code. Dynamic execution is applied as a technique to detect defects and to determine quality attributes of the code. This dynamic testing option is not applicable for the majority of the software work products. Among the questions that arise are:

- How can we evaluate or analyze a work product, such as a requirement specification, a user story, a design document, a test plan or a user manual?
- How can we effectively examine the source code before execution?

As discussed above, one powerful technique that can be used is static testing, for example reviews. In principle, all human-readable software work products can be tested using review techniques.

Types of defects that are easier to find during static testing include:

- Requirements defects, such as inconsistencies, ambiguities, contradictions, omissions, inaccuracies, redundancies.
- Design defects, such as inefficient algorithms or database structures, poor modularization.
- Coding defects, such as variables with undefined values, variables that are declared but never used, unreachable code, duplicate code, excessive code complexity. All of these can be found by static analysis tools.
- Deviations from standards, for example lack of adherence to naming conventions in coding standards.
- Incorrect interface specifications, such as different units of measurement used by the calling system than by the called system or a mismatch of the number, type, or order of parameters. In 1999, a Mars Orbiter burned up in the atmosphere due to one team using metric units of measurement and the other using imperial units of measurement, as described in NASA [1999].
- Security vulnerabilities, such as buffer overflow susceptibility.
- Traceability problems, such as gaps or inaccuracies or lack of coverage (for example, missing tests for an acceptance criterion).
- Maintainability defects, such as improper modularization, poor reusability, code that is difficult to analyze and modify (often referred to as code smells). These defects do not show up in dynamic testing but can be critical for long-term costs of the system.

Compared to dynamic testing, static testing finds defects rather than failures. You may recall that in Chapter 1 Section 1.2.3 we made a distinction between errors, defects and failures. An error is a mistake made by a human being (for example, in writing code), a defect is something that is wrong (for example, in the code itself) and a failure is when the system or component does not perform as it should (for example, returns the wrong balance). Static testing does not cause the system or component to do anything, so it cannot find failures; only dynamic testing can do that. However, static testing does find defects directly. Dynamic testing has to investigate the failure to find a defect.

In addition to finding defects, the objectives of reviews can also be informational, communicational and educational. Participants learn about the content of software work products to help them understand the role of their own work and to plan for future stages of development. Reviews often represent project milestones and support the establishment of a baseline for a software product. The type and quantity of defects found during reviews can also help testers focus their testing and select effective classes of tests. In some cases, customers/users or product owners attend the review meeting and provide feedback to the development team, so reviews are also a means of customer/user communication.

To summarize, the main differences between static and dynamic testing are:

- They find different types of defects.
- Static testing finds defects directly; dynamic testing finds failures (caused by defects).
- Static testing may find defects in code that is seldom executed.
- Static testing applies primarily to non-executable work products and dynamic testing to executable ones.
- They measure different kinds of quality characteristics.

3.2 FEEDBACK AND REVIEW PROCESS

SYLLABUS LEARNING OBJECTIVES FOR 3.2 FEEDBACK AND REVIEW PROCESS (K2)

FL-3.2.1   **Identify the benefits of early and frequent stakeholder feedback (K1)**

FL-3.2.2   **Summarize the activities of the review process (K2)**

FL-3.2.3   **Recall which responsibilities are assigned to the principal roles when performing reviews (K1)**

FL-3.2.4   **Compare and contrast the different review types (K2)**

FL-3.2.5   **Recall the factors that contribute to a successful review (K1)**

In this section, we will focus on reviews as a distinct—and distinctly useful—form of static testing. We'll discuss the process for carrying out reviews. We'll talk about who does what in a review meeting and as part of the review process. We'll cover types of reviews that you can use. We'll look at different variations for reviews based on different ways to prepare for and perform reviews, and finally we will look at success factors to enable the most effective and efficient reviews possible. As we go through this section, watch for the Syllabus terms **anomaly**, **formal review**, **informal review**, **inspection**, **review**, **technical review** and **walkthrough**. You will find these terms also defined in the Glossary.

One reason why reviews are so useful is that having a different person look at a work product is a way to overcome cognitive bias, the tendency to see what we intended rather than what we actually wrote.

Review A type of static testing in which a work product or process is evaluated by one or more individuals to detect defects or to provide improvements.

Informal review A type of review that does not follow a defined process and has no formally documented output.

Formal review A type of review that follows a defined process with a formally documented output.

Reviews vary from very informal to formal (that is, well-structured and regulated). Although inspection is perhaps the most documented and formal review technique, it is certainly not the only one. The formality of a review process is related to factors such as the maturity of the development process, any legal or regulatory requirements or the need for an audit trail. In practice, the **informal review** is perhaps the most common type of review. Informal reviews are applied at various times during the early stages in the life cycle of a work product. A two-person team can conduct an informal review, as the author can ask a colleague to review a work product or code. Pair working (pair programming, pair testing or a tester and developer pairing) is also an informal way to review the work products that both are working on. In later stages, reviews often involve more people and a meeting. **Formal reviews** generally have team (peer) participation, documented results of the review and specified procedures to follow in carrying out the review. They may involve discussions or focus on finding defects. The goal is to help the author and to improve the quality of the work product. Informal reviews come in various shapes and forms, but all have one characteristic in common: they are not documented.

Different reviews may have a different focus or objective for the review. For example, one objective may be to find defects. This is often at least one of the objectives of most types of review, formal or informal. Sometimes the objective is for all participants to gain knowledge and understanding. Although a walkthrough is normally used for this purpose, an informal review could also meet this goal, and it is often a by-product (if not an explicit objective) for technical reviews or inspections. Another objective may be to hold discussions and come to a consensus about technical issues; this is normally the focus of a technical review.

3.2.1 Benefits of early and frequent stakeholder feedback

A feedback loop is where information from a process is fed back into an earlier stage of that process. We have all heard the squeal of a loudspeaker where the sound feeds back into a microphone, in this case making an unpleasant noise. An example of good feedback is when you are driving a car—you make a small adjustment to the steering wheel and observe where you are on the road; your road position indicates what adjustment you should make next to the steering to keep on the road. A slow feedback loop is a boat or ship—an adjustment to the speed or direction is made, but it may be minutes rather than seconds when you know whether you are coming in to dock too fast.

In software development, we want to have early feedback from stakeholders about what features are being developed and how they will work. A stakeholder will communicate a vision of what they want to have, and the developers receive and interpret their requirements. But since no communication (especially between humans) is error-free, what the developers produce may not be what the stakeholder actually needs. The earlier we find this out, the more efficiently we can produce the end product that will be what is actually needed. If we don't have early feedback from stakeholders, we risk delivering the wrong thing, resulting in costly rework, missed deadlines, blame games, or even complete failure of the whole project.

The feedback also needs to be frequent—imagine if you could only adjust your road position once a minute—you might be alright on a long straight main road, but not on a winding country road. Frequent communication with stakeholders and feedback about what is being developed can prevent misunderstandings about what is wanted or needed, particularly when requirements may change over time through the increased understanding of the stakeholders and the developers. This can enable developers to focus on producing what will give the most value most quickly to the stakeholders and will reduce risk.

3.2.2 Review process activities

The standard that covers review processes is ISO/IEC 20246 [2017]. The standard defines a generic review process which can be tailored for each situation. More formal reviews follow a formal process, which may explicitly include the separate activities below. Informal reviews may perform at least some of the same activities to some extent, but not necessarily as separate activities.

In this section, we describe good review practices in more detail than is required by the Syllabus, because we know that doing reviews well brings great benefits.

The review process consists of a number of main activities that we discuss next.

Planning

The following are the elements of the planning activity:

- Defining the scope of the review, for example the purpose of the review, what work products (for example, documents) or parts of work products to review and the quality characteristics to be evaluated in the review.
- Deciding what areas to focus on in the review.
- Identifying review characteristics such as the type of review with roles, activities and checklists, as well as supporting information needed such as standards.
- Defining the exit criteria for more formal review types (for example, inspections).
- Estimating effort and the timeframe for the review.
- Selecting the people to participate in the review and allocating roles to each reviewer.
- Checking that entry criteria are met before the review starts (for more formal review types).

Note that the last two are not specifically mentioned in the Syllabus but would be part of formal review planning.

Let's examine these in more detail.

The review process for a particular review may begin with a request for review by the author to the review leader, who takes overall responsibility for the review, for example scheduling (dates, time, place and invitation) of the review. On a project level, the project planning needs to allow time for review and rework activities, thus providing engineers with time to thoroughly participate in reviews.

For more formal reviews, for example inspections, the moderator (facilitator) performs an entry check and defines at this stage formal exit criteria. The entry check is carried out to ensure that the reviewers' time is not wasted on a work product that is not ready for review. A work product containing too many obvious mistakes is clearly not ready to enter a formal review process, and it could even be very harmful to the review process. It would possibly de-motivate both reviewers and the author. Also, the review is likely to be less effective because the numerous obvious and minor defects will conceal the major defects.

The following are possible entry criteria for a formal review:

● A short check of a work product sample by the review leader (or expert) does not reveal many major defects.

● The work product to be reviewed is available with line numbers (if relevant).

● The work product has been cleaned up by running any applicable automated checks, such as static analysis or spelling and grammar assessments.

● References needed for the review are stable and available.

● The work product author is prepared to join the review team and feels confident with the quality of the product.

If the work product passes the entry check, the review leader and author decide which part(s) of it to review. Because the human mind can comprehend a limited set of pages at one time, the number should not be too high. The maximum number of pages depends, among other things, on the objective, review type and work product type. It should be derived from practical experiences within the organization. For a review, the maximum size is usually between 10 and 20 pages. In formal inspection, only a page or two may be looked at in depth (comparing to related documents) in order to find the most serious defects that are not obvious.

After the work product size has been set and the pages to be checked have been selected, the review leader determines, in co-operation with the author, the composition of the review team. The team normally consists of three to six participants, including moderator (facilitator) and author. To improve the effectiveness of the review, different roles may be assigned to each of the participants. These roles help the reviewers focus on particular types of defects during individual review. This reduces the chance of different reviewers finding the same defects. The review leader or moderator assigns the roles to the reviewers.

Quality characteristics may also be evaluated and documented in a review, for example the testability of a design or the readability or understandability of user help or installation instructions. These may also be assigned roles.

Review initiation

The goal of this activity is to ensure that everyone who will be involved in the review knows what is expected, and that everything that they will use in the review is ready. The following are the activities for initiating a review:

- Ensuring access to the work products (physically or electronically) and any other relevant material such as logging forms, checklists or related work products.

- Explaining the scope, objectives, process, roles and work products to the participants.

- Answering any questions that participants may have about the review.

Let's examine these in more detail.

The goal of this set of activities is to get everybody on the same wavelength regarding the work product under review and to commit to the time that will be spent on checking (that is, individual reviewing). The result of the entry check and defined exit criteria are discussed in case of a more formal review. This stage of the review process is important to increase the motivation of reviewers and thus the effectiveness of the review process. At customer sites, we have measured results of up to 70% more major defects found per page as a result of performing a kick-off meeting, a form of review initiation, as described in van Veenendaal and van der Zwan [2000].

The review initiation may be done remotely, or it may involve a meeting (in person or using video conferencing). If a meeting is held, the reviewers receive a short introduction to the objectives of the review and the work products. The relationships between the work product under review and the other work products (for example, sources or predecessor work products) are explained, especially if the number of related work products is high.

Role assignments, checking rate, the pages to be reviewed, the roles to be taken by each person, process changes and possible other questions are also discussed during this meeting.

Whether or not a meeting is held, the moderator (facilitator) ensures that each reviewer is clear about their responsibilities, and answers any questions that they may have, either about the work products or the review process itself.

Individual review (also called individual preparation)

The following are the activities for the individual review:

- Reviewing all or part of the work documents(s).

- Logging any **anomaly**, defect, recommendation or question, possibly applying one or more review techniques such as checklist-based reviewing or scenario-based reviewing.

> **Anomaly** A condition that deviates from expectation.

Let us examine these in more detail.

In the individual review, the participants work alone on the work product under review using the related work products, procedures, rules and checklists provided. The individual participants identify **anomalies** (potential defects), recommendations, questions and comments, according to their understanding of the work product and the particular role they have been given. All issues are recorded, preferably using a logging form. Spelling mistakes may be recorded on the work product under review, but not mentioned during the meeting. The annotated work product may be given or sent to the author at the end of the logging meeting. Using checklists during individual reviewing can make reviews more effective and efficient, for example a specific checklist based on perspectives such as user, maintainer, tester or operations, or a checklist for typical coding problems. This may also be an assigned role. More details can be found in ISO/IEC 20246 [2017].

Note that the term 'anomaly' (or 'issue') is a preferable term to use rather than to ask reviewers to find 'defects'. When something is noticed in a work product, it may or may not be a defect. There might be an inconsistency between the work product and a related document, but the reviewer doesn't know which product is correct and which is wrong. It is also less threatening to the author to record anomalies rather than defects.

A critical success factor for a thorough preparation is the number of pages individually reviewed per hour. This is called the checking rate. The optimum checking rate is the result of a mix of factors, including the type of work product, its complexity, the number of related work products and the experience of the reviewer. Usually, the checking rate is in the range of five to ten pages per hour, but may be much less for formal inspection, for example one page per hour. This rate is not staring at a single page for an hour but is comparing in detail the single (critically important) page to a number of other documents and standards. During preparation, participants should not exceed the checking rate they have been asked to use. By collecting data and measuring the review process, company-specific criteria for checking rate and work product size (refer to Planning) can be set, preferably specific to a work product type.

Communication and analysis

The following are the activities for communication and analysis:

- Communicating identified anomalies (potential defects), for example in a review meeting.
- Analyzing anomalies, assigning ownership and status to them.
- Evaluating and documenting quality characteristics of the work product.
- Evaluating the review findings against the exit criteria to make a review decision (reject; major changes needed; accept, possibly with minor changes, or possibly with a follow-up review).

Let's examine these in more detail.

In a formal review, the things found by the individual reviewers are communicated to the author of the work product (or the person who will fix the ones that turn out to be real defects) and may also be communicated to the other reviewers. We refer to these as anomalies at this point because we do not yet know if they are defects or not. The anomalies may be communicated electronically, or in a review meeting, which may consist of the following activities (partly depending on the review type): logging, discussion and decision making.

Note that the Syllabus says that 'all anomalies need to be analyzed and discussed', but this is actually not the case. They do all need to be analyzed, but they do not all need to be discussed—in fact this can be very wasteful of both author's and reviewer's time. Some review types do encourage discussion, but inspection actually keeps discussion to a minimum.

Logging in a review meeting

During or before logging, the anomalies that have been identified during the individual review are collected together. These previously found anomalies may just be given to the author to deal with, or some or all of them may be mentioned in the review meeting, before looking for more anomalies. A separate person to do the logging in the meeting (a scribe) is especially useful for formal review types such as an inspection. To ensure progress and efficiency, discussion is minimized during logging. If an anomaly needs discussion, the item is noted as a discussion item and then handled in the later discussion part of the meeting. A detailed discussion on whether or not an anomaly is a defect is not very meaningful, as it is much more efficient to simply log it and proceed to the next one. Furthermore, in spite of the opinion of the team, a discussed and discarded anomaly may well turn out to be a real defect during rework.

Every anomaly identified as a "real" defect should be logged, along with its severity. The participant who identifies the defect may propose the severity, or the moderator may assign a severity. If reviewers assign severity, it is important that the meaning of each category is understood by all reviewers in the same way. Otherwise, one reviewer may regard everything as critical, for example, skewing the review results. Severity classes could be:

- *Critical*: defects will cause downstream damage; the scope and impact of the defect is beyond the work product under inspection.
- *Major*: defects could cause a downstream effect (for example, a fault in a design can result in an error in the implementation).
- *Minor*: defects are not likely to cause downstream damage (for example, non-compliance with the standards and templates).

In order to keep the added value of reviews, cosmetic issues or minor spelling errors are not part of the defect classification. Such defects are noted by the participants in the work product under review and given to the author at the end of the meeting, or could be dealt with in a separate proofreading exercise.

During logging, the focus is on logging as many defects as possible within a certain timeframe. To ensure this, the moderator tries to keep a good logging rate (number of defects logged per minute). In a well-led and disciplined formal review meeting, the logging rate should be between one and two defects logged per minute.

Discussion part of a review meeting

For a more formal review, the issues classified as discussion items will be handled during a discussion part of the review meeting, which occurs after the logging has been completed. Less formal reviews will often not have separate logging and discussion parts and will start immediately with logging mixed with discussion (which may lead to fewer defects being found). Participants can take part in the discussion by bringing forward their comments and reasoning. In the discussion part of the meeting, the moderator takes care of people issues. For example, they prevent discussions from getting too personal, rephrase remarks if necessary and call for a break to cool down heated discussions and/or participants.

Reviewers who do not need to be in the discussion may leave, or may stay as a learning exercise. The moderator also paces this part of the meeting and ensures that all discussed items either have an outcome by the end of the meeting or are noted as an action point if the issue cannot be solved during the meeting. The outcome of discussions is documented for future reference.

Quality characteristics may also be evaluated and documented at this point, for example the testability of a design or the readability or understandability of user help or installation instructions.

Decision-making part of a review meeting

At the end of the meeting, a decision on the work product under review has to be made by the participants, sometimes based on formal exit criteria. The most important exit criterion may be the average number of critical and/or major defects found per page (for example, no more than three critical/major defects per page). If the number of defects found per page exceeds a certain level, the work product may need to be reviewed again, after it has been reworked. If the work product complies with the exit criteria, it will be checked later by the review leader, moderator or one or more participants. Subsequently, the work product can leave the review process.

In addition to the number of defects per page, other exit criteria are used that measure the thoroughness of the review process, such as ensuring that all pages have been checked at the right rate. The average number of defects per page is only a valid quality indicator if these process criteria are met.

If a project is under pressure, the review leader or moderator will sometimes be forced to skip re-reviews and exit with a defect-prone work product. Setting (and agreeing to) quantified exit criteria helps the review leader or moderator to make firm decisions at all times. Even if a limited sample of a work product has been (formally) reviewed, an estimate of remaining defects per page can give an indication of likely problems later on.

For informal reviews, some of these activities may be performed, but informally. For example, potential defects may be emailed to the author of the work product with a suggested severity classification. The author may then evaluate exit criteria, possibly checking back with the reviewer(s).

Fixing and reporting

The following are the activities for fixing and reporting:

- A defect report is created for those findings that require changes, so that corrective actions can be followed up.

- Checking that exit criteria are met (for more formal review types) so that the work product can be accepted.

- Gathering metrics (for more formal review types), for example of defects fixed, deferred, etc. to form the report for the review results.

Let's examine these in more detail.

Defect reports may have been recorded in a general defect logging tool (for example, also used by testers) or may have been recorded in a review log as anomalies.

During fixing, the author will improve the work product under review step by step, based on the anomalies or defects detected by the individual reviewers and/or those found in the review meeting. Not every anomaly that is reported is a defect that leads to a fix. It is often the author's responsibility to judge if an anomaly really is a defect which has to be fixed, though in some cases the review meeting participants may make those decisions. If nothing is done about an anomaly for a certain reason, it should still be reported to show that the author has considered it and why they have rejected it as a defect.

Changes that are made to the work product should be easy to identify by anyone who will be confirming that the fixes are correct. For example, the author may turn on 'track changes' in a document.

Sometimes an anomaly raised affects a work product that is not under the direct control of the author of the reviewed work product. For example, reviewing a feature to be implemented may reveal an inconsistency or omission in the user story or requirements specification that it is based on. If the author cannot update the user story or requirement specification directly, they need to communicate this to whoever can update the related work product.

As defects are fixed, the status of those defects should be updated wherever they are managed. In some reviews, the originator of the comment or defect would then confirm that the defect has been fixed adequately, that the author has correctly understood their comment and has fixed the right problem in the right way.

In order to control and optimize a formal review process, a number of measurements are collected by the moderator at each step of the process. Examples of such measurements include number of defects found, number of defects found per page, time spent checking per page, total review effort, etc. It is the responsibility of the moderator to ensure that the information is correct and stored for future analysis.

The moderator is responsible for ensuring that satisfactory actions have been taken on all logged defects, process improvement suggestions and change requests. Although the moderator checks to make sure that the author has taken action on all known defects, it is not necessary for them to check all the corrections in detail. If it is decided that all participants will check the updated work product, the moderator takes care of the distribution and collects the feedback. For more formal review types the review leader or moderator checks for compliance to the exit criteria.

When all the exit criteria have been met, including fixing, checking of fixes and confirming that the review process was properly carried out, then the work product can be formally accepted. This may be particularly important in safety-critical systems.

3.2.3 Roles and responsibilities in a formal review

The participants in any type of formal review should have adequate knowledge of the review process. The best and most efficient review situation occurs when the participants gain some kind of advantage for their own work during reviewing. In the case of an inspection or technical review, participants should have been properly trained, as both types of review have proved to be far less successful without trained participants. This indeed is a critical success factor.

The best formal reviews come from well-organized teams, guided by trained moderators (facilitators). Within a review team, six types of roles can be distinguished: manager, author, moderator (or facilitator), scribe (or recorder), reviewers and review leader.

Manager
Managers have a number of very important responsibilities in successful reviews, including:

- Ensuring that reviews are planned and what should be reviewed.
- Providing resources such as staff, budget and time.

The role of managers in reviews is often underestimated, but without adequate support from managers, reviews are seldom successful. The manager needs to believe in reviews and to ensure that they will be carried out as part of development. This includes allowing time in schedules for reviews to be done (and also any resulting rework!). Managers should also keep an eye on the effectiveness of reviews and encourage increasing effectiveness and efficiency, so that the greatest benefits are gained from reviews (not specifically mentioned in the Syllabus but important for successful reviews). The manager should have clear objectives for the review process(es) and should determine whether those objectives have been met. The manager will ensure that any review training requested by the participants takes place. Of course, a manager can also be involved in the review itself, depending on their background, playing the role of a review leader or reviewer if this would be helpful. In some review types, the manager should not be involved, for example as moderator of the review meeting (for example, inspection).

The author

The author has two main responsibilities:

- Creating the work product under review.
- Fixing defects in the work product (if necessary).

The author's basic goal should be to learn as much as possible with regard to improving the quality of the work product, but also to improve their ability to write future work products. The author's task is to illuminate unclear areas and to understand the defects found.

Moderator (also known as facilitator)

The responsibilities of the moderator are:

- Ensuring the effective running of review meetings (when held).
- Mediating, if necessary, between the various points of view.
- Time management of the review meeting.
- Ensuring a psychologically safe environment (blame-free) in which reviewers can speak freely and the focus is on learning.

The moderator is the person upon whom the success of the review often depends. They lead each individual review meeting and the associated activities. The moderator performs the entry check and checks on the fixes, in order to control the quality of the input and output of the review process. The moderator also schedules the meeting, disseminates work products and other relevant materials, organizes the meeting, coaches other team members, paces the meeting, leads possible discussions and stores the data that is collected. While the moderator can do their best to conduct the meeting in such a way as to encourage a safe working environment, the broader corporate culture will have a greater effect on whether this is even possible. A strong culture of learning from mistakes will mean that reviews are safe places to raise anomalies; a blame culture will encourage hiding of mistakes and suppression of the lessons that could improve everyone's work products (and job satisfaction).

Scribe (also known as recorder)

The responsibilities of the scribe or recorder are:

- Collating anomalies found during the individual review activity.
- Recording new anomalies, open points and decisions from the review meeting (when held).

During the logging meeting, the scribe (or recorder) has to record each anomaly mentioned and any suggestions for process improvement. In practice, it is often the author who plays the role of scribe, ensuring that the log is readable and understandable. If authors record their own defects, or at least make their own notes in their own words, it helps them to understand the log better when they look through it afterwards to fix the defects found. However, having someone other than the author taking the role of the scribe (for example, the moderator) can have significant advantages, since the author is freed up to think about the work product rather than being tied down with lots of writing.

The role of the scribe or recorder may be less relevant, or even irrelevant, if the defects, discussion points, decisions and other issues raised in the review are recorded electronically, although the scribe may be the person doing that recording during a meeting.

Reviewers

The responsibilities of the reviewers are:

- Being subject matter experts, persons working on the project, stakeholders with an interest in the work product and/or individuals with specific technical or business backgrounds.
- Identifying anomalies (potential defects) in the work product under review.

The role of the reviewers (also called checkers or inspectors) is to be another set of eyes to look at the work products from a fresh point of view. The reviewers check any material for potential defects, mostly prior to the meeting. The level of thoroughness required depends on the type of review. The level of domain knowledge or technical expertise needed by the reviewers also depends on the type of review. Reviewers should be chosen to represent different perspectives in the review process. In addition to the work product under review, reviewers also receive other material, including source work products, standards, checklists, etc. In general, the fewer source and reference work products provided, the more domain expertise is needed regarding the content of the work product under review.

Review leader

The responsibilities of the review leader are:

- Taking overall responsibility for the review.
- Deciding who will be involved.
- Organizing when and where the review will take place.

Depending on the size of the organization, the role of the review leader may be taken by a manager or by the moderator. If review leader is a separate role, they will be working closely with both of these roles.

In some organizations, there is no distinction made between the review leader and the moderator in practice. The main difference is that the review leader is responsible for the review happening and organizes the people involved, but they may not be involved in the review meeting (if this is how issues are communicated). The moderator's main role is in dealing with the people while the review is happening, ensuring that the review meetings are run well and making sure that interpersonal issues do not disrupt the review process.

Although we have described the roles and responsibilities as though they are all separate and distinct, this does not mean that they are done by different people. One person may take on more than one role and perform the responsibilities for multiple roles. The determination of who does what role also depends on the type of review.

3.2.4 Types of review

The different review types have different objectives and can be used for different purposes, but the most common objective of all review types is to uncover potential defects in the work product being reviewed. The type of review should be chosen, based on a number of factors, including the SDLC being followed, the maturity of the development process, needs of the project, available resources, criticality and complexity of the product type and risks, legal or regulatory requirements, the need for an audit trail, the business domain and company culture. The objectives of the review (refer to Section 3.2.5) are key for choosing the right kind(s) of review for each situation and context.

The main review types, their main characteristics and attributes are described below.

Informal review (for example, buddy check, pairing, pair review)

An informal review is characterized by the following attributes:

- Main purpose/objective: detecting anomalies (potential defects).
- Not based on a defined process.

An informal review may simply be one person saying to a colleague, 'Could you have a quick look at what I've just done?' The colleague may spend less than an hour looking through and giving any comments back to the author, such as typos, something missing, or a 'But have you thought of this?' comment. With the right person reviewing, this buddy check can be very effective (and at little cost in time). However, this form of review is dependent on the individual doing the review; the greater their expertise, the more useful the review will be.

Other forms of informal review include pair working, when one person works with another to produce a work product, the second person continually evaluating (that is, reviewing) what the first person is typing. A checklist can be used to increase the effectiveness of an informal review. Because they generally give faster feedback, informal reviews can be used a number of times during development.

Walkthrough

Walkthrough A type of review in which an author leads members of the review through a work product and the members ask questions and make comments about possible issues.

A **walkthrough** is characterized by the following attributes:

- Main purposes: evaluating quality, building confidence in the work product, educating reviewers, gaining consensus, generating new ideas, motivating and enabling authors to improve, and detecting anomalies.
- Individual preparation before the review meeting is optional.
- Review meeting is typically led by the author of the work product.

Within a walkthrough, the author does most of the preparation. The participants, who are selected from different departments and backgrounds, are not generally required to do a detailed study of the work products in advance (but it is an option). Because of the way the meeting is structured, a large number of people can participate, and this larger audience can bring a great number of diverse viewpoints regarding the contents of the work product being reviewed. If the audience represents a broad cross-section of skills and disciplines, it can give assurance that no major defects are missed in the walkthrough. A walkthrough is especially useful for higher-level work products, such as requirement specifications and architectural documents.

A walkthrough is often used to transfer knowledge and educate a wider audience about a particular work product. In some cases, the educational value of a walkthrough is more important than finding defects (although defects should be welcomed).

If a large number of people are present, a scribe (someone other than the author) may be used to record the discussion, the questions raised and any decisions taken at the meeting.

Technical review

A **technical review** is characterized by the following attributes:

- Main purposes: gaining consensus, making decisions about technical problems, detecting anomalies, evaluating quality and building confidence in the work product, generating new ideas and motivating and enabling authors to improve.
- Reviewers should be technical peers of the author, and technical experts in relevant disciplines.
- Review meeting is led by a trained moderator (typically not the author).

A technical review is often a discussion meeting that focuses on achieving consensus about the technical content of a work product that all the participants have studied before the meeting. During technical reviews, anomalies are found by experts, who focus on the content of the work product. The experts who participate in a technical review may include, for example, architects, chief designers and key users. It is useful to have an independent moderator, especially if there are a number of strong opinions about technical issues. Technical reviews are typically less formal than inspections but generally more formal than walkthroughs. In practice, technical reviews may vary from quite informal to very formal.

> **Technical review**
> A formal review by technical experts that examine the quality of a work product and identify discrepancies from specifications and standards.

Inspection

An **inspection** is characterized by the following attributes:

- The full defined process is followed (refer to Section 3.2.2), with formal documented outputs, based on rules and checklists.
- Main objectives are to detect as many anomalies as possible.
- Other objectives include evaluating quality and building confidence in the work product, motivating and enabling authors to improve future work products.
- The review meeting is led by a trained moderator (not the author).
- The author cannot act as the review leader, moderator, reader or scribe.
- Metrics are collected and used to improve the entire software development process, including the inspection process.

Inspection is the most formal review type. The work product under inspection is prepared and checked thoroughly by the reviewers before the meeting, comparing the work product with its sources and other referenced work products, and using rules and checklists. In the inspection meeting, the defects found are logged and any discussion is postponed until the discussion part of the meeting. This makes the inspection meeting a very efficient meeting.

Depending on the organization and the objectives of a project, inspections can be balanced to serve a number of goals. For example, if the time to market is extremely important, the emphasis in inspections will be on efficiency. In a safety-critical market, the focus will be on effectiveness.

> **Inspection** A type of formal review that uses defined team roles and measurement to identify defects in a work product, and improve the review process and the software development process.

When inspections are done well, they not only help to identify defects in the work products being reviewed, but the emphasis on process improvement and learning leads to better ways of producing work products, both for the author and for other reviewers.

Figure 3.1 shows some different roles with respect to documents or work products used within a formal review, particularly inspection. The roles represent views of the work product under review:

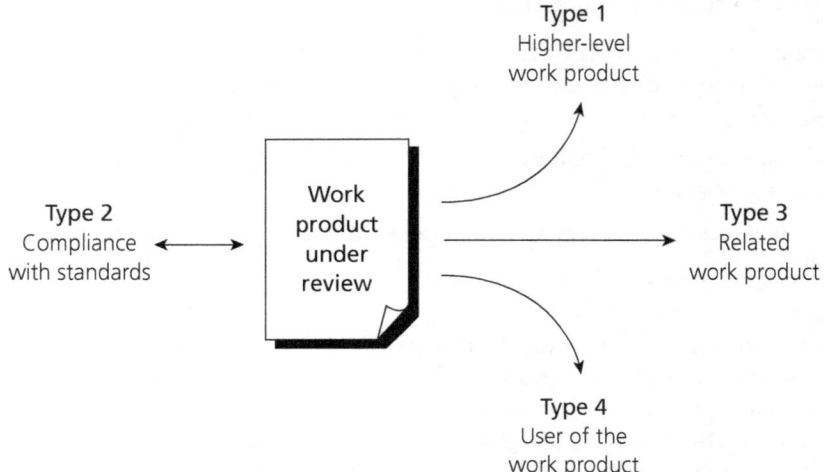

FIGURE 3.1   Basic review roles for a work product under review

- Focus on higher-level work products, for example does the design comply with the requirements (Type 1 in Figure 3.1, also known as 'source' documents).
- Focus on standards, for example internal consistency, clarity, naming conventions, templates (Type 2).
- Focus on related work products at the same level, for example interfaces between software functions (Type 3).
- Focus on usage of the work product, for example for testability or maintainability (Type 4).

All review types

A single work product may be the subject of more than one review. If more than one type of review is used, the order may vary. For example, an informal review may be carried out before a technical review to make sure the work product is ready for the technical review, or an inspection may be carried out on a requirements specification or epic before a walkthrough with customers. No one of the types of review is the winner, but the different types serve different purposes at different stages in the life cycle of a work product.

A peer review is where all of the reviewers are at the same or similar organizational level as the author, that is, the author's equals are reviewing the work. A peer review is not another type of review, as all of the types of review could be carried out by peers, from informal review to inspection.

All review types have as at least one of their purposes to find anomalies. The most important thing is to find the most severe defects, those that represent the highest risk to the organization. Reviewing the most important work products is one way to get greater value from reviews. For example, a decimal point error may not be very important in a report about sales of low-volume items, but a decimal point error in a multi-million dollar or euro contract could be very significant. The types of defect found depend on the work product, as well as on the reviewers and the way in which the reviews are carried out.

There are a number of specific ways in which reviews can be carried out, which were covered in more detail in the previous Syllabus. We include descriptions of them here as additional information (not on the exam) which we believe is useful.

■ ■ ■ ■

The previous Syllabus included information about other ways that reviews can be made more effective. We include this additional information here in this 'optional extra' section, as it is useful for improving the way reviews are carried out.

Checklist-based reviewing

Checklist-based reviewing is often far more effective than just an ad hoc review, because there is helpful guidance given about what to look for in the supplied checklist. The checklist would be one of the things distributed when the review is initiated. Different reviewers may have different checklists. This helps to cut down the number of duplicate defects found by reviewers. A checklist may contain items to check or questions.

The checklist may contain questions relating to the work product, such as 'Have references been given for claims shown?', 'Are all pointers valid?' or 'Has the customer's viewpoint been considered?' It is a good idea to organize a checklist so that if the answer to a question is no then there is a potential defect. The best checklist questions are often derived from something that went wrong or was missed in a previous review. If a major defect was missed, adding a specific question to a checklist to look for that type of defect is very effective.

However, checklists should not just be allowed to grow and grow. It is important to review the checklists regularly and to remove questions that are no longer useful, so that the checklist is focused on the most important things. The longer a checklist is, the less likely it is to be fully used, so it is also a good idea to limit the size of the checklist to one page and to put the most important checklist questions at the beginning.

Checklists are useful when reviewing any type of work product, from unit or component code to user stories or Help text. The questions would, of course, be different for each work product, as they are targeted at potential defects for that type of work product. For example, code-specific questions (about pointers) would be appropriate for a unit code checklist, and customer-related questions would be appropriate for a user story or requirements specification checklist.

When checklists are used well, this is a very effective review technique; if all of the checklist questions are evaluated, this is a systematic search for previous or typical defect types.

One last word of warning about checklist-based reviews: do not be constrained to check only what is on the checklist. The checklist can help to guide you to look for specific things, but also be aware of other things, and take inspiration from the checklist to look for related potential defects outside of the checklist.

Scenario-based reviewing and dry runs

A scenario-based review is one where the reviewers are given a structured perspective on how to work through a work product from a particular point of view. For example, if use cases are used, a use case can be a scenario to work through how the system will work from each actor's point of view (people and other systems).

When you are stepping through the scenario, this is also called a 'dry run', a term used for a rehearsal before a big event or speech. In this sense, going through the scenario is a rehearsal for the system. We are looking at the way the system is expected to be used, which is an important aspect of validation.

A scenario is a different way of reviewing from using a checklist. You are looking at how the system will be used from a specific perspective, and this can be more focused than just a list of questions or points to check. When using a scenario, we are acting out realistic ways of using the system and validating whether or not it will meet user needs and expectations. A checklist is more likely to be focused on verifying individual aspects, such as types of defect, that have occurred previously.

As with checklists, do not be constrained by the scenario, but use it as inspiration for finding other defects outside of the scenario. With user-focused scenarios, be particularly aware of missing features.

Role-based reviewing

Role-based reviewing is similar to scenario-based reviewing, but the viewpoints are different stakeholders rather than just users. Roles can include specific end-user types, such as experienced versus inexperienced, age-related (children, teenagers, adults, seniors), or accessibility roles such as vision impaired, hearing impaired, etc.

Personas may also be used, which are typically based on a profile of a specific set of characteristics. A persona is the characterization of a user who represents the target audience for your system. A number of different personas are used to represent a range of user profiles, needs or desires. For example, one persona may be a young adult who is single and likes skiing; another may be a married elderly person with significant health problems. Each persona is described in detail (job, where they live, income level, etc.). Each persona would represent a role in role-based reviewing.

The author may raise additional specific roles and questions to be addressed. The moderator has the option to also fulfil a role, alongside the task of being the facilitator for the meeting. Checking the work product improves the moderator's ability to lead the meeting, because it ensures better understanding. Furthermore, it improves the review efficiency because the moderator replaces an engineer who would otherwise have to check the work product and attend the meeting. It is recommended that the moderator take the role of checking compliance to standards, since this tends to be a highly objective role, with less discussion about the defects found.

Roles can also be organizational, such as from the perspective of a system administrator or user administrator. They can also be from testing perspectives such as performance testing or security testing.

In inspection, viewpoint roles can also include reviewing from the perspective of contractual issues, manufacturing (if relevant), legal issues, design or testing, operations or delivery or a third-party supplier. In addition to these viewpoint roles, there can be procedural roles, such as checking all financial calculations, starting from the back, looking for the most important things or cross-checking within or between work products. There may also be document or work product roles where each reviewer is given special responsibility for using one particular related work product. For example, if there are three user stories that are related to the work product being reviewed, and there are three reviewers, each would pay special attention to one of the three user stories.

Perspective-based reading

Perspective-based reading is similar to role-based reviewing, but rather than playing a specific role, the reviewer typically tries to perform the tasks on a high level that they would be doing with the work product under review (taking that perspective), for example as a tester making some test designs. Stakeholder perspectives include end-users, marketing, designer, tester or operations. These are similar to some of the inspection roles, and in fact this technique was devised for inspections. This technique goes beyond role-based and is therefore typically more expensive, but it also finds more defects.

The benefits of this technique (and also role-based and scenario-based reviewing) are that each individual reviewer has their own specialty within the review and will be looking in more depth within their own assigned area. This makes the review more effective (since there is deeper checking) but also more efficient, since there is less duplication of issues found by different reviewers. Checklists can also be used for the different perspectives.

Another approach to perspective-based reading is for the reviewers to use the work product to generate other work products from it. When doing perspective-based reading on a requirements specification, a tester-perspective reviewer would be generating acceptance tests. This can help to identify missing information needed for the tests.

Perspective-based reading combines aspects of all of the review techniques (except *ad hoc*) and so can be the most effective. There is no one right way to do reviews, but when reviewers look at the work product in different ways, using checklists, scenarios, roles or perspectives, then the best results are gained. More information is available in Shull, Rus and Basili [2000].

3.2.5 Success factors for reviews

Implementing (formal) reviews is not easy. There is no one way to success and there are numerous ways to fail. The most common reasons for failure in reviews are due to either organizational factors or people-related factors. We will look at aspects of both of these.

One aspect (which is not emphasized in the Syllabus) is the importance of having a champion, the person who will lead the process on a project or organizational level. They need expertise, enthusiasm and a practical mindset in order to guide managers, review leaders, moderators and participants. The authority of this champion should be clear to the entire organization.

To be successful, a review culture should be part of the mindset of the organization, and there are a number of aspects which can make or break the effectiveness and efficiency of reviews within an organization. Here are some of the main success factors for reviews.

Define clear objectives and measurable exit criteria

Each review should have clear objectives and measurable exit criteria (especially formal reviews). These are defined during the planning stage, communicated to all participants and may be used to decide the outcome of the review or the definition of done for the review.

It is particularly important that the review does not in any way attempt to evaluate the individual performance of the author—this is the way to successfully kill the review process! A successful review has an atmosphere of trust and learning, where mistakes are not a trigger for blame but a real learning opportunity. This must be clear as an objective for the reviews in order to gain real benefit from them.

Choose the appropriate review type and technique

In the previous subsections, we discussed a number of review types and techniques, each of which has its strengths and weaknesses, and advantages and disadvantages in use. You should be careful to select and use review types and techniques that will best enable the achievement of the objectives of the project and the review itself.

Be sure to consider the type, importance and risk level of the work product to be reviewed and the reviewers who will participate. For example, do not try to review everything by inspection; fit the review to the risk associated with specific parts of the work product. Some work products may only warrant an informal review and others will repay using inspection. Of course, it is also of utmost importance that the right people are involved. Consider the work product to be reviewed. Would a checklist-based review technique or a role-based technique be the most suitable for identifying defects? Which review type would be best for the needs of the project, type of product and the context of development and review processes?

In addition, it is also important to choose the right reviewers (though not emphasized in the Syllabus). The people chosen to participate in a review have a significant impact on the success and outcome of the review. Different review types or techniques may require different skills from the reviewers, and different types of work product may require different skills. For example, if code is being reviewed, the reviewers must at least be able to read and understand the code, so developer skills are needed.

If role-based review techniques are used, people who have or can adopt those particular perspectives are needed as reviewers. If someone will be using a particular work product in their own work (for example, a developer who will be working to implement a feature in a user story), that person would be a good candidate to include on the review team, as they have a vested interest in understanding their source (the user story), clarifying any ambiguities and removing any defects before they use it in their own development work.

Provide feedback to stakeholders and authors

Feedback should be provided for stakeholders so that they are aware of progress and are alerted to any major quality concerns as soon as possible (refer also to Section 3.2.1). Report quantified and aggregated results and benefits at a high (not individual) level to all those involved as soon as possible, including discussion of the consequences of defects if they had not been found this early. Costs should of course be tracked, but benefits, especially when problems do not occur in the future, should be made visible by quantifying the benefits as well as the costs.

It is important to give feedback to the authors while they are still working on the work product. Early and frequent feedback is more effective and more efficient, partly because if a particular type of defect is detected early, the author can be aware of this in the rest of the work product; this prevents that type of defect from appearing in the parts of the work product written later.

Provide adequate time especially for preparation

Reviews take time, and this time needs to be scheduled. But it is not just the time for review meetings. It is more important to allow adequate time for reviewers to do the preparation and individual studying of the work product before any meeting, as this is when most defects are identified. Skimping on preparation time is a false economy. The review meetings must be scheduled with adequate notice, so that reviewers can plan the review time into their own work schedules. It is not realistic to expect reviewers to be able to drop everything for a review; they need to balance their own work with the review work. Don't forget time for fixing defects found!

Management support is critical

Management support is essential for success. Managers should, among other things, incorporate adequate time for review activities in project schedules. They should visibly support the review process and help to foster a culture where reviews are seen as valuable both to the organization and to the individual. Managers especially must commit not to use metrics or other information from reviews for the evaluation of the author or the participants. If people are blamed for making mistakes, they do not make fewer mistakes, they hide them better!

One of the authors was called in to give training for reviews in an organization where the manager claimed to be very much in favour of using them. This sounded good, but during the training, it became clear that something was not right; there was great resistance to the idea of reviews. When raised with the manager, he said, 'But I am really supporting reviews here—that way I can find out who puts in the most defects and fire them'. Needless to say, reviews did not work in this organization!

To ensure that reviews become part of the day-to-day activities, the hours to be spent should be made visible within each project plan. The engineers involved should be prompted to schedule time for preparation and, very importantly, rework. Tracking these hours will improve planning of the next review. As stated earlier, management plays an important part in the planning of review activities.

Make reviews part of the culture

Reviews flourish within organizations that have a learning and open culture, and they die in organizations with a blame culture.

A culture of learning paves the way for process improvement, not only in the processes producing the work products being reviewed but also improvements in the review process itself.

The attitude towards defects found in reviews is critical to success. Any defect found is an opportunity to improve not only the quality of the work product that it was found in but also to become aware of other similar defects in the same or in different work products. When reviews are working well, many more defects are prevented than are found.

But the attitude towards defects by the author of the work product is also critical. Defects should be acknowledged and appreciated (even if you are not sure at the time whether it really is a defect or not, acknowledge the point made by the person reporting it). Defects should be reported and handled objectively; defect reporting should never become personal, as this creates ill will and damages the review process and culture.

The way to tell what sort of culture your organization is, is to ask: 'What question is asked when something goes wrong?' If the answer is 'Who's to blame?', reviews are unlikely to succeed. If the answer is 'What can we all learn from this?', you are in a good organization and reviews are likely to be a great help.

Provide adequate training

It is important that training is provided in review techniques, especially the more formal techniques, such as inspection. Otherwise, the process is likely to be impeded by those who do not understand the process and the reasoning behind it. Special training should be provided to the moderators (facilitators) to prepare them for their critical role in the review process.

Each reviewer has responsibilities and possibly roles to play in the review. It is important that each of them takes the reviewing work seriously and does it to the best of their ability. This includes spending adequate time on the review activities, and paying attention to detail as needed, for example by using a checklist. Good training will help every reviewer to do the reviewing task better.

Review meetings are facilitated and well managed

The role of the moderator or facilitator in the review meeting is one of the major factors in making the whole review experience useful and pleasant. The meeting should be focused on the review objectives, and any discussion should be tightly controlled. It is all too easy for discussions in meetings to go on and on, but if the purpose of the review is to identify defects as efficiently as possible, then most discussion is probably unnecessary. All review participants should feel that their time in the review (including the meeting) has been a valuable use of their time.

Reviews are about evaluating someone's work product. Some reviewers tend to get too personal when they are not well managed by the moderator (facilitator). People issues and psychological aspects should be dealt with by the moderator and should be part of the review training, thus making the review a positive experience for the author. During the review, defects should be welcomed. This is much more likely when they are expressed objectively. It is important that all participants create and operate in an atmosphere of trust.

It is very important that the defect information is considered sensitive data and handled accordingly. Even the rumour of a manager wanting to use review data to evaluate people is enough to destroy the effectiveness of a review process. There needs to be an atmosphere of trust among the reviewers; they all know that they are helping each other get better, and they need to be confident that the data will be used in the right way.

Particularly in review meetings, subliminal factors such as body language and tone of voice may damage the openness of a good review atmosphere. In addition, if defects raised are criticized, for example by the author, then reviewers are much less likely to raise other defects, so something very important may be missed. Language is important as well. Contrast 'There may be a problem here' (objective wording), with 'You did it wrong here' (personal attack).

Additional factors

We have listed just some of the factors that contribute to successful reviews, mainly focusing on organizational factors. However, there are also some other factors worth mentioning.

Testers make good reviewers. As discussed in Chapter 1, testers are professional pessimists. This focus on what could go wrong makes them good contributors to reviews, provided they observe the earlier points about keeping the review experience a positive one. In addition to providing valuable input to the review itself, testers who participate in reviews often learn about the product. This supports earlier testing, one of the principles discussed in Chapter 1. Using testers as reviewers is also beneficial to the testers, as the review can help them identify relevant test conditions and test cases, and to begin preparing the tests earlier.

Do not try to review too much at one time. Even if a limited scope has not been formally defined, reviewers should be selective about what they spend their time on. Select the work products for review that are most important in a project. Reviewing highly critical, upstream work products like requirements and architecture will most certainly show the benefits of the review process to the project. This investment in review hours will have a clear and high return on investment. Another advantage of limiting the scope of a review is that it is easier to concentrate on a small piece of work rather than trying to take in a large amount. Reviewers need to keep up their concentration on the important things, both in individual preparation and in any review meeting

We use work products that support the review process to perform a review. These work products need to be up to date and of good quality to support the reviews. Review materials, such as checklists, need reviewing too! When significant defects are found in work products (either in reviews or in testing), add an item to the relevant checklist so that this type of work product defect can be identified earlier next time. At regular intervals, check to make sure that all of the items or questions on a checklist are still relevant, and remove any that are not useful anymore.

Finally, the review process is simple but not easy. Every step of the process is clear, but experience is needed to execute them correctly. Try to get experienced people to observe and help where possible. But most importantly, start doing reviews and start learning from every review.

More information about successful reviews can be found in Weigers [2002], van Veenendaal [2004], Sauer [2000] and Kramer and Legeard [2016].

CHAPTER REVIEW

Let's review what you have learned in this chapter.

From Section 3.1, you should be able to recognize the software work products that can be examined by static testing techniques. You should be able to explain the value of static testing by using examples. You should be able to explain the difference between static testing and dynamic testing in terms of their objectives, types of defects to be found and the role of these techniques within the software life cycle. You should know the Glossary terms **dynamic testing**, **static analysis** and **static testing**.

From Section 3.2, you should be able to summarize the activities of the work product feedback and review process, and the benefits of early and frequent stakeholder feedback. You should be able to summarize the different activities of review processes. You should recognize the different roles and responsibilities in a formal review. You should be able to explain the differences between the various types of review: informal review, walkthrough, technical review and inspection. Finally, you should be able to explain the factors for successful performance of reviews, both organizational and people-related. You should know the Glossary terms **anomaly**, **formal review**, **informal review**, **inspection**, **review**, **technical review** and **walkthrough**.

SAMPLE EXAM QUESTIONS

Question 1 Which of the following artefacts can NOT be examined using review techniques?

a. Software code.

b. User story.

c. Test designs.

d. User's intentions.

Question 2 Which of the following are the main activities of the work product review process?

1. Planning.

2. Review initiation.

3. Select reviewers.

4. Individual review.

5. Review meeting.

6. Evaluating review findings against exit criteria.

7. Communication and analysis.

8. Fixing and reporting.

a. 1, 2, 4, 7, 8.

b. 2, 3, 4, 5, 8.

c. 1, 2, 3, 5, 7.

d. 1, 4, 5, 6, 7.

Question 3 Which statement about static and dynamic testing is True?

a. Static testing and dynamic testing have different objectives.

b. Static testing and dynamic testing find the same types of defect.

c. Static testing identifies defects through failures; dynamic testing finds defects directly.

d. Static testing can find some types of defect with less effort than dynamic testing.

Question 4 Which statement about reviews is True?

a. Inspections are led by a trained moderator, but this is not necessary for technical reviews.

b. Technical reviews are led by a trained leader, inspections are not.

c. In a walkthrough, the author does not attend.

d. Participants for a walkthrough always need to be thoroughly trained.

Question 5 Which statement below is True?

a. Management ensures effective running of review meetings; the review leader decides who will be involved.

b. Management is responsible for review planning; the moderator monitors ongoing cost effectiveness.

c. Management organizes when and where reviews will take place; the review leader assigns staff, budget and time.

d. Management decides on the execution of reviews; the moderator is often the person on whom the success of the review depends.

Question 6 Match the following characteristics with the type of review.

1. Led by the author.

2. Undocumented.

3. Reviewers are technical peers of the author.

4. Led by a trained moderator or leader.

5. Uses entry and exit criteria.

INSP: Inspection

TR: Technical review

IR: Informal review

W: Walkthrough

a. INSP: 4, TR: 3, IR: 2 and 5, W: 1

b. INSP: 4 and 5, TR: 3, IR: 2, W: 1

c. INSP: 1 and 5, TR: 3, IR: 2, W: 4

d. INSP: 5, TR: 4, IR: 3, W: 1 and 2

Question 7 Which of the following statements about success factors in reviews are True?

1. Reviewers should try to review as much of the work product as they can.

2. The author acknowledges and appreciates defects found in their work.

3. Each review has clear objectives which are communicated to the reviewers.

4. Testers are not normally involved in reviews, as their work focuses on test design.

5. Support from managers is not necessary, since they will not be reviewers.

a. 2 and 3.

b. 1, 2 and 3.

c. 1, 2 and 4.

d. 2, 3 and 5.

Question 8 Which of the following is an effect of early and frequent stakeholder feedback?

a. Deadlines are guaranteed to be met.

b. The feedback will show that the stakeholders need to be more thorough in documenting their requirements formally.

c. Early communication of potential quality problems may avoid or lessen their effect.

d. Early and frequent communication will make it clear that the developers' view of the stakeholder requirements should be implemented.

Question 9 Why can static testing be more efficient and less costly than dynamic testing?

a. Defects found in reviews are more important, static analysis tools are cheap.

b. Defects found in reviews can only be fixed in testing, static analysis only works on code.

c. Defects found in reviews can be fixed earlier, static analysis tools find defects quickly.

d. Defects found in reviews don't need to be fixed, defects found by static analysis are fixed by the tool.

EXERCISE

Perform an individual review of the functional specification shown in Document 3.1, using the checklist below:

1. Do all requirements give sufficient detail needed for determining the expected results for tests?
2. Are all dependencies and restrictions specified?
3. Are alternatives specified for all options?

Make a list of potential defects in the specification, noting the severity level of each (High, Medium or Low). Note which checklist question has found the defect. Solution ideas are given in the next section.

DOCUMENT 3.1 Functional requirements specification

Functional requirements

This specification describes the required functionality of the booking system for a sports centre.

Browsing the facilities

Customers will be able to view all the available facilities using the following options to filter the selection:

- none: shows all facilities
- outdoor/indoor
- sport: selected from a pull-down list
- date: using a standard calendar
- time: selected from a pull-down list.

Each facility shown will include colour-coded information on availability (not available, already booked and available for booking) and the time slots.

Selecting a facility

Any facility available for booking can be selected by clicking on the required time period. A confirmation message will be displayed. The user may either cancel the message and return to the list of facilities or continue to the booking details page.

Booking details

If the customer is not already logged in, the customer login dialogue will be displayed.

Details of the facility together with any relevant regulations (such as age restrictions, footwear and no-show conditions, etc.) will be displayed. The customer name, member ID and mobile phone number fields will be pre-filled. The customer can either cancel or confirm the details. Confirming the details takes the customer to the payment details page; cancelling returns the customer to the list of facilities page.

Payment details

Payment can be made by credit card. The following details are required:

- card type: Visa or Mastercard
- name on card: as printed on card
- card number: 16 digits, no spaces
- expiry date: mm/yy
- CVC code: last three digits on the signature strip.

Having entered these details, the system will seek authorization for the payment. When this is received a final confirmation message is displayed. The customer can choose to cancel or confirm. If confirmed, the booking is made and a confirmation email is sent to the customer's email address (as specified in their customer profile). If the payment authorization fails, an error message is displayed.

EXERCISE SOLUTION

Table 3.1 lists some potential defects in the functional requirement shown in Document 3.1. This is not an exhaustive list; you may have found others. We have noted the checklist question number used to discover each issue, and a severity level of High, Medium or Low.

TABLE 3.1   Potential defects in the functional requirements specification

Defect	Description	Checklist	Severity
1	Need to know the specific colours in the colour coding to be able to see if the test passes or fails.	1	M
2	Need to know what all the facilities are in order to check if they are displayed correctly.	1	M
3	What facilities are in the 'Sport' pull-down list?	2	M
4	What are the rules for selection of options? Does Date have to be selected before Time, for example? Can you select only one option?	2	H
5	If login fails, what is supposed to happen: can you continue to book anyway?	3	H
6	If login is successful, what screen is displayed: already selected or back to list?	3	L
7	Can filters be combined? For example, can you select Indoor and Tennis?	2	L
8	Is it possible to choose more than one time slot at once, for example 3 consecutive half-hour times?	2	M
9	If credit card authorization fails, what happens after the error message is shown?	3	L
10	Can the customer cancel after authorization has gone through? Has payment already been taken? If so, will it be refunded? Automatically?	3	H
11	When a booking is cancelled after authorization, what screen is shown after the error message?	3	L
12	Missing from specification: is cancellation possible after a booking has been made, either by the customer or by the sports centre? Is a refund given? Automatically?	none	H

As mentioned above, these are some suggested potential defects. Note that we have phrased many of them as questions, which may be less threatening to the author of the requirements document. Note also that we have noted an additional high severity defect which was not specifically triggered by an item in the checklist.

CHAPTER FOUR

Test analysis and design

Chapter 3 covered static testing, looking at work products and code, but not running the code we are interested in. This chapter looks at dynamic testing, where the software we are interested in is run by executing tests on the running code.

4.1 TEST TECHNIQUES OVERVIEW

SYLLABUS LEARNING OBJECTIVE FOR 4.1 TEST TECHNIQUES OVERVIEW (K2)

FL-4.1.1   Distinguish black-box, white-box and experience-based test techniques (K2)

In this section we will look at the different types of test techniques, how they are used, how they differ and the factors to consider when choosing a test technique. The three types or categories are distinguished by their primary source: a description of what the system should do (for example, requirements, user stories, etc.), the structure of the system or component, or a person's experience. All categories are useful and the three are complementary.

The purpose of a test technique is to identify test conditions, test cases, test data and coverage items. Test conditions are identified during analysis and then used to define test cases and test data during test design, which are then used in test implementation. Coverage items may be chosen in either test analysis or design. A way to remember the difference is that test analysis helps us decide <u>what</u> to test; test design and test implementation helps us decide <u>how</u> to test. For example, in risk-based testing strategies, we identify risk items (which are the test conditions) when performing an analysis of the risks to product quality. Those risk items, along with their corresponding levels of risk, are subsequently used to design the test cases and implement the test data. Risk-based testing will be discussed in Chapter 5.

Test techniques also help us to devise sets of tests that are both effective (e.g. at finding defects) and efficient (taking the minimum number of tests to find them).

In this section, look for the definitions of the Glossary terms **black-box test technique**, **coverage**, **coverage item**, **experience-based test technique**, **test technique** and **white-box test technique**.

There are many different types of software **test technique**, each with its own strengths and weaknesses. Each individual technique is good at finding particular types of defect and relatively poor at finding other types. For example, a technique

Test technique A procedure used to define test conditions, design test cases, and specify test data. (Also known as test case design technique, test specification technique, test design technique).

that explores the upper and lower limits of a single input range is more likely to find boundary value defects than those associated with combinations of inputs. Similarly, testing performed at different stages in the software development life cycle will find different types of defects; component testing is more likely to find coding logic defects than system design defects or user experience problems.

Each test technique falls into one of a number of different categories. Broadly speaking there are two main categories: static and dynamic. Static test techniques, as discussed in Chapter 3, do not execute the code being examined and are generally used before any tests are executed on the software. They could be called non-execution techniques. Most static test techniques can be used to test any form of work product including source code, design documents and models, user stories, functional specifications and requirement specifications. Static analysis is a tool-supported type of static testing that concentrates on testing formal languages and so is most often used to statically test source code.

In this chapter we look at dynamic test techniques, which are subdivided into three more categories: black-box (also known as specification-based, behavioural or behaviour-based techniques), white-box (structure-based or structural techniques) and experience-based. Black-box test techniques include both functional and non-functional techniques (that is, testing of quality characteristics). The techniques covered in the Syllabus are summarized in Figure 4.1.

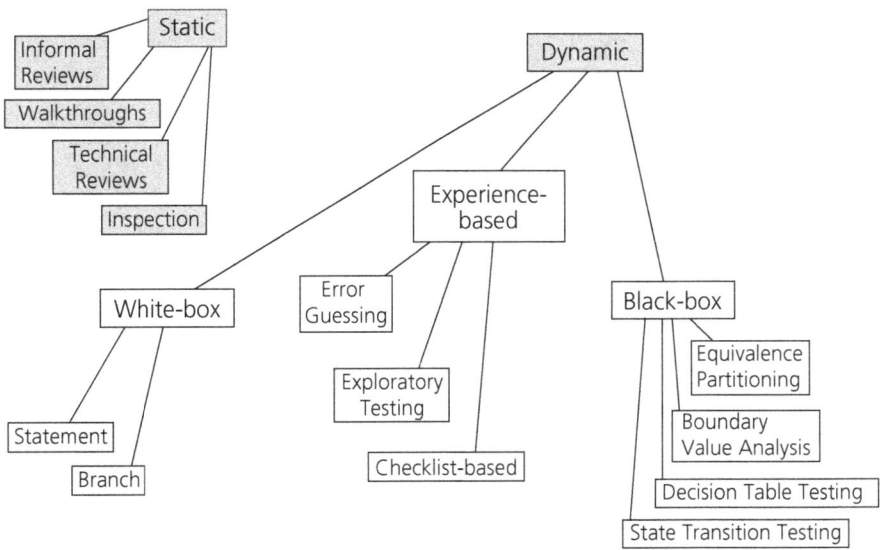

FIGURE 4.1   Test techniques

4.1.1 Black-box, white-box and experience-based test techniques

Black-box test techniques

The first of the dynamic test techniques we will look at are the **black-box test techniques**. They are called black-box because they view the software as a black box with inputs and outputs, but they have no knowledge of how the system or component is structured inside the box. In essence, the tester is concentrating on what the software does, not how it does it.

Black-box test technique A test technique based on an analysis of the specification of a component or system.

All black-box test techniques have the common characteristic that they are based on a model (formal or informal) of some aspect of the system, which enables test conditions and test cases to be derived from them in a systematic way.

Common characteristics of black-box test techniques include the following:

- Test conditions, test cases and test data are derived from a test basis that may include software requirements, specifications, use cases and user stories. The source of information for black-box tests is some description of what the system or software is supposed to do.

- Test cases may be used to detect gaps between the requirements and the implementation of the requirements, as well as deviations from the requirements. One of the strengths of test cases is that they make things specific, and this often highlights different understandings about the test basis, showing what is missing or interpreted differently.

- Coverage is measured based on the items tested in the test basis and the technique applied to the test basis. As we will see later, whenever you can make a list of some things that could be tested and can tell whether they have been tested, then you can measure coverage. Coverage at black-box level is based on items from the test basis. For example, does every requirement described have at least one test that exercises it?

Black-box test techniques are appropriate at all levels of testing (component testing through to acceptance testing) where a specification or other test basis exists. When performing system or acceptance testing, the requirements specification or functional specification may form the basis of the tests. When performing component or integration testing, a design document or low-level specification may form the basis of the tests.

Black-box techniques include both functional and non-functional testing. Functional testing is concerned with what the system does, its features or functions. Non-functional testing is concerned with examining how well the system does something, rather than what it does. Non-functional aspects (also known as quality characteristics or quality attributes) include performance, usability, portability, maintainability, etc. Techniques to test these non-functional aspects are less procedural and less formalized than those of other categories, as the actual tests are more dependent on the type of system, what it does and the resources available for the tests.

Non-functional testing is part of the Syllabus and is also covered in Chapter 2. There are techniques for deriving non-functional tests [Gilb 1988], but they are not covered at the Foundation level.

Because the black-box test cases are independent of how the system is implemented, if the implementation changes but the required behaviour stays the same, then the tests are still useful and valid. An example of this is where existing regression tests should still pass even if part of the system has been upgraded.

Categorizing test techniques into black-box and white-box is mentioned in a number of testing books, including Myers *et al.* [2011], Beizer [1990], Black [2007], Copeland [2004], Craig *et al.* [2002], Koomen *et al.* [2006], Jorgensen [2014], Ammann [2016] and Forgács and Kovács [2019].

White-box test technique A test technique only based on the internal structure of a component or system.

White-box test techniques

White-box test techniques (which are also dynamic rather than static) use the internal structure of the software to derive test cases. They are commonly called white-box or glass-box techniques (implying you can see into the system/box) since they require knowledge of how the software is implemented, that is, how it works. For example, a structural technique may be concerned with exercising loops in the software. Different test cases may be derived to exercise the loop once, twice and many times. This may be done regardless of the functionality of the software. All structure-based techniques have the common characteristic that they are based on how the software under test is constructed or designed. This structural information is used to assess which parts of the software have been exercised by a set of tests (often derived by other techniques). Additional test cases can then be derived in a systematic way to cover the parts of the structure that have not been touched by any test before.

Note that the definition of white-box techniques is referring to the way in which test inputs are derived, to measure or exercise a particular structure. However, a test case is more than its inputs, and the expected results cannot be derived from the structure—they must come from some documented or undocumented test oracle external to the software itself. Otherwise, the test is simply telling you that the software does what the software does—what we want to know is, does the software do what the software should do!

Common characteristics of white-box test techniques include the following:

- Test conditions, test cases and test data are derived from a test basis that may include code, software architecture, detailed design or any other source of information regarding the structure of the software. White-box test techniques are most commonly used for code structure, but these techniques are also useful for other structures. For example, the menu structure of an app could be tested using white-box techniques.

- **Coverage** is measured based on the **coverage items** tested within a selected structure, for example the code statements, the decisions, the interfaces, the menu structure or any other identified structural element. Refer to Section 4.3 for more on coverage.

- White-box test techniques determine the path through the software that either was taken or that you want to be taken, and this is determined by specific inputs to the software. However, in order to be a test, we also need to know what the expected outcome of the test case should be, even though this does not affect the path taken. A test oracle of some kind, for example a specification, is used to determine the expected outcome.

White-box test techniques can also be used at all levels of testing. Developers use white-box techniques in component testing and component integration testing, especially where there is good tool support for code coverage. White-box techniques are also used in system and acceptance testing, but the structures are different. For example, the coverage of major business transactions could be the structural element in system or acceptance testing.

Because they are based on a structure, white-box tests can only be derived after that structure exists, so after the test object has been created. In contrast, black-box tests can (and should) be created before the test object has been built.

Experience-based test techniques

In **experience-based test techniques**, people's knowledge, skills and background are a prime contributor to the test conditions, test cases and test data. The experience of both technical and business people is important, as they bring different perspectives to the test analysis and design process. Due to previous experience with similar systems,

Coverage (test coverage) The degree to which specified coverage items are exercised by a test suite, expressed as a percentage.

Coverage item An attribute or combination of attributes derived from one or more test conditions by using a test technique.

Experience-based test technique A test technique based on the tester's experience, knowledge and intuition.

they may have insights into what could go wrong, which is very useful for testing. Both structural and behavioural insights are used to design experience-based tests.

All experience-based test techniques have the common characteristic that they are based on human knowledge and experience, both of the system itself (including the knowledge of users and stakeholders) and of likely defects. Test cases are therefore derived in a less systematic way but may be more effective.

Experience-based test techniques are used to complement black-box and white-box techniques, and are also used when there is no specification, or if the specification is inadequate or out-of-date. This may be the only type of technique used for low-risk systems, but this approach may be particularly useful under extreme time pressure. In fact this is one of the factors leading to exploratory testing.

4.2 BLACK-BOX TEST TECHNIQUES

> **SYLLABUS LEARNING OBJECTIVES FOR 4.2 BLACK-BOX TEST TECHNIQUES (K3)**
>
> **FL-4.2.1** Use equivalence partitioning to derive test cases (K3)
>
> **FL-4.2.2** Use boundary value analysis to derive test cases (K3)
>
> **FL-4.2.3** Use decision table testing to derive test cases (K3)
>
> **FL-4.2.4** Use state transition testing to derive test cases (K3)

In this section we will look in detail at four black-box test techniques. These four techniques are K3 in the Syllabus. This means that you need to be able to use these techniques to design test cases. In Section 4.3, we will look at the K2 white-box test techniques, and in Section 4.4, the K2 experience-based techniques. We end the chapter with a look at collaboration-based test approaches, which can be used with any of the techniques.

In this section, look for the definitions of the Glossary terms **boundary value analysis**, **decision table testing**, **equivalence partitioning** and **state transition testing**.

The four black-box test techniques we will cover in detail are:

- equivalence partitioning
- boundary value analysis
- decision table testing
- state transition testing.

4.2.1 Equivalence partitioning

Equivalence partitioning A black-box test technique in which test conditions are equivalence partitions exercised by one representative member of each partition.

Equivalence partitioning (EP) is a good all-round black-box test technique. It can be applied at any level of testing and is often a good technique to use first. It is a common-sense approach to testing, so much so that most testers practise it informally even though they may not realize it. However, while it is better to use the technique informally than not at all, it is much better to use the technique in a formal way to attain the full benefits that it can deliver. This technique will be found in most testing books (refer to the list in Section 4.1 and References near the end of this book.)

The idea behind the technique is to divide (that is, to partition) a set of test conditions into groups or sets where all elements of the set can be considered the same, so the system should handle them equivalently, hence 'equivalence partitioning'. **Equivalence partitions** are also known as equivalence classes: the two terms mean exactly the same thing. Note that equivalence partitions must not overlap and must not be empty. They can be continuous (e.g. decimal numbers between 10 and 20) or discrete (e.g. the colours of the Olympic rings), ordered (e.g. numbers) or unordered (e.g. colours), finite (e.g. age between 17 and 25) or infinite (e.g. numbers greater than 15). The individual elements of a partition may be in more than one partition, for example the number 19 is in all of the example partitions above except for those relating to colours.

> **Equivalence partition** A subset of the value domain of a variable within a component or system in which all values are expected to be treated the same based on the specification.

The EP technique then requires that we need test only one condition from each partition. This is because we are assuming that all the conditions in one partition will be treated in the same way by the software. If one condition in a partition works, we assume all of the conditions in that partition will work, and so there is little point in testing any of these others. Conversely, if one of the conditions in a partition does not work, then we assume that none of the conditions in that partition will work, so again there is little point in testing any more in that partition. Of course, these are simplifying assumptions that may not always be right, but if we write them down, at least it gives other people the chance to challenge the assumptions we have made and hopefully help to identify better partitions. If you have time, you may want to try more than one value from a partition, especially if you want to confirm a selection of typical user inputs.

For example, a savings account in a bank earns a different rate of interest depending on the balance in the account. In order to test the software that calculates the interest due, we can identify the ranges of balance values that earn the different rates of interest. For example, if a balance in the range $0 up to $100 has a 3% interest rate, a balance over $100 and up to $1,000 has a 5% interest rate, and balances of $1,000 and over have a 7% interest rate, we would initially identify three valid equivalence partitions and one invalid partition as shown below:

Invalid partition	Valid (for 3% interest)		Valid (for 5%)		Valid (for 7%)
−$0.01	$0.00	$100.00	$100.01	$999.99	$1,000.00

Notice that we have identified four partitions here, even though the specification only mentions three. This illustrates a very important task of the tester: not only do we test what is in our specification, but we also think about things that have not been specified. In this case we have thought of the situation where the balance is less than zero. We have not (yet) identified an invalid partition on the right, but this would also be a good thing to consider. To identify where the 7% partition ends, we would need to know what the maximum balance is for this account (which may not be easy to find out). In our example we have left this open for the time being. Note that non-numeric input is also an invalid partition (for example, the letter 'a') but we discuss only the numeric partitions for now.

We have made an assumption here about what the smallest difference is between two values. We have assumed two decimal places, that is $100.00, but we could have assumed zero decimal places (that is $100) or more than two decimal places (for example $100.0000). In any case it is a good idea to state your assumptions, then other people can see them and let you know if they are correct or not.

We have also made an assumption about exactly which amount starts the new interest rate: a cent over to go into the 5% interest rate, but exactly on the $1,000.00 to go into the 7% rate. By making our assumptions explicit, and documenting them in this technique, we may highlight any differences in understanding exactly what the specification means.

When designing the test cases for this software we would ensure that the three valid equivalence partitions are each covered once, and we would also test the invalid partition at least once. So, for example, we might choose to calculate the interest on balances of −$10.00, $50.00, $260.00 and $1,348.00. If we had not specifically identified these partitions, it is possible that at least one of them could have been missed at the expense of testing another one several times over. Note that we could also apply EP to outputs as well. In this case we have three interest rates: 3%, 5% and 7%, plus the error message for the invalid partition (or partitions). In this example, the output partitions line up exactly with the input partitions.

How would someone test this without thinking about the partitions? A naïve tester (let's call him Robbie) might have thought that a good set of tests would be to test every $50. That would give the following tests: $50.00, $100.00, $150.00, $200.00, $250.00... say up to $800.00 (then Robbie would have got tired of it and thought that enough tests had been carried out). But look at what Robbie has tested: only two out of four partitions! So, if the system does not correctly handle a negative balance or a balance of $1,000 or more, he would not have found these defects, so the naïve approach is less effective than EP. At the same time, Robbie has four times more tests (16 tests versus our four tests using equivalence partitions), so he is also much less efficient! This is why we say that using techniques such as this makes testing both more effective and more efficient.

Note that when we say a partition is invalid, it does not mean that it represents a value that cannot be entered by a user or a value that the user is not supposed to enter. It just means that it is not one of the expected inputs for this particular field. The software should correctly handle values from the invalid partition, by replying with an error message such as 'Balance must be at least $0.00'.

Note also that the invalid partition may be invalid only in the context of crediting interest payments. An account that is overdrawn will require some different action.

Here is a summary of EP characteristics:

- Valid values should be accepted by the component or system. An equivalence partition containing valid values is called a valid equivalence partition.

- Invalid values should be rejected by the component or system. An equivalence partition containing invalid values is called an invalid equivalence partition.

- Partitions can be identified for any data element related to the test object, including inputs, outputs, internal values, time-related values (for example, before or after an event) and for interface parameters (for example, integrated components being tested during integration testing).

- Any partition may be divided into sub-partitions if required, where smaller differences of behaviour are defined or possible. For example, if a valid input range goes from –100 to 100, then we could have three sub-partitions: valid and negative, valid and zero, and valid and positive.

● Each value belongs to one and only one equivalence partition from a set of partitions. However, it is possible to apply EP more than once and end up with different sets of partitions, as we will see later under Section 4.2.2.

● When values from valid partitions are used in test cases, they can be combined with other valid values in the same test, as the whole set should pass. We can therefore test many valid values at the same time.

● When values from invalid partitions are used in test cases, they should be tested individually, that is, not combined with other invalid equivalence partitions, to ensure that failures are not masked. Failures can be masked when several failures occur at the same time but only one is visible, causing the other failures to be undetected.

● EP is applicable at all test levels.

Coverage is a measure of how thoroughly you have applied a particular test technique, as described in Section 4.1. For equivalence partitions, the coverage items are the partitions we have identified. If we have a test for each partition, then we have 100% EP coverage.

Test objects often contain multiple input parameters, for example a form which has fields for name, address, etc. Every test needs to have something for each field (and as already pointed out, one test can have lots of valid partition values). 'Each choice coverage' requires one value from each set of partitions but does not consider combinations of partitions. For example, if a password needs to be between 6 and 10 characters and must include a capital letter, number and one of the special characters '@ % &', then we need a test that includes each of these, such as 'Xc@b569'.

We have more things to say about EP, but we will return to that after we cover boundary value analysis, since the two techniques are closely related.

4.2.2 Boundary value analysis

Boundary value analysis (BVA) is based on testing at the boundaries between partitions that are ordered, such as a field with numerical input or an alphabetical list of values in a menu. It is essentially an enhancement or extension of EP and can also be used to extend other black-box (and white-box) test techniques. If you have ever done 'range checking', you were probably using the BVA technique, even if you were not aware of it. Note that we have both valid boundaries (in the valid partitions) and invalid boundaries (in the invalid partitions).

> **Boundary value analysis** A black-box test technique in which test cases are designed based on boundary values.

As an example, consider a printer that has an input option of the number of copies to be made, from 1 to 99.

Invalid	Valid	Invalid
0	1 99	100

To apply BVA, we will take the minimum and maximum (boundary) values from the valid partition (1 and 99 in this case) together with the first or last value respectively in each of the invalid partitions adjacent to the valid partition (0 and

100 in this case). In this example we would have three EP tests (one from each of the three partitions) and four boundary value tests.

Let's return to the savings account system described in the previous section:

Invalid partition	Valid (for 3% interest)	Valid (for 5%)	Valid (for 7%)
−$0.01	$0.00 $100.00	$100.01 $999.99	$1,000.00

Because the boundary values are defined as those values on the edge of a partition, we have identified the following boundary values: −$0.01 (an invalid boundary value because it is at the edge of an invalid partition), $0.00, $100.00, $100.01, $999.99 and $1,000.00, all valid boundary values.

So by applying BVA we will have six tests for boundary values. Compare what our naïve tester Robbie had done: he did actually hit one of the boundary values ($100) though it was more by accident than design. So in addition to testing only half of the partitions, Robbie has only tested one-sixth of the boundaries (so he will be less effective at finding any boundary defects). If we consider all our tests for both EP and BVA, the techniques give us a total of ten tests, compared to the 16 that Robbie had, so we are still considerably more efficient as well as being over three times more effective (testing four partitions and six boundaries, so ten conditions in total compared to three).

Note that in this savings account interest example, we have valid partitions next to other valid partitions. If we were to consider an invalid boundary for the 3% interest rate, we have −$0.01, but what about the value just above $100.00? The value of $100.01 is not an *invalid* boundary; it is actually a *valid* boundary because it falls into a valid partition. So the partition for 5%, for example, has no invalid boundary values associated with partitions next to it.

A good way to represent the valid and invalid partitions and boundaries is in a table such as Table 4.1.

TABLE 4.1 Equivalence partitions and boundaries

Test conditions	Valid partitions	Invalid partitions	Valid boundaries	Invalid boundaries
Balance in account	$0.00 − $100.00	< $0.00	$0.00	− $0.01
	$100.01 − $999.99	> $Max	$100.00	$Max + 0.01
	$1,000.00 − $Max	non-integer (if balance is an input field)	$100.01	
			$999.99	
			$1,000.00	
			$Max	
Interest rates	3%	Any other value	Not applicable	Not applicable
	5%	Non-integer		
	7%	No interest calculated		

By showing the values in the table, we can see that no maximum has been specified for the 7% interest rate. We would now want to know what the maximum value is for an account balance, so that we can test that boundary. This is called an open boundary, because one of the sides of the partition is left open, that is, not defined. But that doesn't mean we can ignore it. We should still try to test it, but how? We have called it $Max to remind ourselves to investigate this.

Open boundaries are more difficult to test, but there are ways to approach them. Actually the best solution to the problem is to find out what the boundary should be specified as! One approach is to go back to the specification to see if a maximum has been stated somewhere else for a balance amount. If so, then we know what our boundary value is. Another approach might be to investigate other related areas of the system. For example, the field that holds the account balance figure may be only six figures plus two decimal figures. This would give a maximum account balance of $999,999.99 so we could use that as our maximum boundary value. If we really cannot find anything about what this boundary should be, then we probably need to use an intuitive or experience-based approach to probe various large values trying to make it fail.

We could also try to find out about the lower open boundary. What is the lowest negative balance? Although we have omitted this from our example, setting it out in the table shows that we have omitted it, so helps us be more thorough if we wanted to be.

Representing the partitions and boundaries in a table such as this also makes it easier to see whether or not you have tested each one (if that is your objective). To achieve 100% coverage for EP, we need to ensure that there is at least one test for each identified equivalence partition. To achieve 100% coverage of boundary values, we need to ensure that there is at least one test for each boundary value identified. Of course, just one test for either a partition or a boundary may not be sufficient testing, but it does show some degree of thoroughness since we have not left any partition or boundary untested.

BVA can be applied at all test levels.

Extending equivalence partitioning and boundary value analysis

So far, by using EP and BVA we have identified conditions that could be tested, that is, partitions and boundary values. The techniques are used to identify test conditions, which could be at a fairly high level (for example, a low interest account) or at a detailed level (for example, a value of $100.00). We have been looking at applying these techniques to ranges of numbers. However, we can also apply the techniques to other things.

Applying to more than numbers

Assume that you are booking a flight where you may have a choice of Economy/ Coach, Premium Economy, Business or First Class tickets. Each of these is an equivalence partition in its own right and should be tested, but it does not make sense to talk about boundaries for this type of partition, which is a collection of valid things. The invalid partition would be an attempt to type in any other type of flight class (for example, Staff). If this field is implemented using a drop-down list, then it should not be possible to type anything else in, but it is still a good test to try at least once in some drop-down field. When you are analyzing the test basis (for example, a requirements specification or user story), EP can help to identify where a drop-down list would be appropriate.

When trying to identify a defect, you might try several values in a partition. If this results in different behaviour where you expected it to be the same, then there may be two (or more) partitions where you initially thought there was only one.

We can apply EP and BVA to all levels of testing. The examples here were at a fairly detailed level, probably most appropriate in component testing or in the detailed testing of a single screen.

At a system level, for example, we may have three basic configurations which our users can choose from when setting up their systems, with a number of options for each configuration. The basic configurations could be system administrator, manager and customer liaison. These represent three equivalence partitions that could be tested. We could have serious problems if we forget to test the configuration for the system administrator, for example.

Applying more than once

We can also apply EP and BVA more than once to the same specification item. For example, if an internal telephone system for a company with 200 telephones has three-digit extension numbers from 100 to 699, we can identify the following partitions and boundaries:

- Digits (characters 0 to 9) with the invalid partition containing non-digits.

- Number of digits, 3 (so invalid boundary values of two digits and four digits).

- Range of extension numbers, 100 to 699 (so invalid boundary values of 099 and 700).

- Extensions that are in use and those that are not (two valid partitions, no boundaries).

- The lowest and highest extension numbers that are in use could also be used as boundary values.

One test case could test more than one of these partitions/boundaries. For example, extension 409 which is in use would test four valid partitions: digits, the number of digits, the valid range and the in-use partition. It also tests the boundary values for digits of 0 and 9.

How many test cases would we need to test all of these partitions and boundaries, both valid and invalid? We would need a non-digit, a two-digit and a four-digit number, the values of 99, 100, 699 and 700, one extension that is not in use, and possibly the lowest and highest extensions in use. This is 10 or 11 test cases: the exact number would depend on what we could combine in one test case.

Using EP and BVA helps us to identify tests that are most likely to find defects and to use fewer test cases to find them. This is because the contents of a partition are representative of all of the possible values. Rather than test all ten individual digits, we test one in the middle (for example, 4) and the two edges (0 and 9). Instead of testing every possible non-digit character, one character can represent all of them.

Applying to output

As we mentioned earlier, we can also apply these techniques to output partitions. Consider the following extension to our bank interest rate example. Suppose that a customer with more than one account can have an extra 1% interest on this account if they have at least $1,000 in this one. Now we have two possible output values (7% interest and 8% interest) for the same account balance, so we have identified another test condition (8% interest rate). (We may also have identified that same output condition by looking at customers with more than one account, which is a partition of types of customer.)

Applying to more than human inputs

EP can be applied to different types of input as well. Our examples have concentrated on inputs that would be typed in (by a human) when using the system. However, systems receive input data from other sources as well, such as from other systems via some interface. This is also a good place to look for partitions (and boundaries). For example, the value of an interface parameter may fall into valid and invalid equivalence partitions. This type of defect is often difficult to find in testing once the interfaces have been joined together, so is particularly useful to apply in integration testing, either component integration, for example between APIs (Application Programming Interfaces) or system integration. If you are receiving data from a third-party supplier, a value sent by the supplier may be larger than the maximum value your system is expecting. If you do not check the value when it arrives, this could cause problems.

Boundary value analysis can be applied to a whole string of characters (for example, a name or address). The number of characters in the string is a partition. So, for example, between 1 and 30 characters is the valid partition with valid boundaries of 1 and 30. The invalid boundaries would therefore be zero characters (null, just hit the Return key) and 31 characters. Both of these should produce an error message.

Partitions can also be identified when setting up test data. If there are different types of record, your testing will be more representative if you include a data record of each type. The size of a record is also a partition with boundaries, so we could include maximum and minimum size records in the test database.

If you have some inside knowledge about how the data is physically organized, you may be able to identify some hidden boundaries. For example, if an overflow storage block is used when more than 255 characters are entered into a field, the boundary value tests would include 255 and 256 characters in that field. This may be verging on white-box testing, since we have some knowledge of how the data is structured, but it does not matter how we classify things as long as our testing is effective at finding defects. Don't get hung up on a fine distinction: just do whatever testing makes sense, based on what you know. An old Chinese proverb says, 'It doesn't matter whether the cat is white or black; all that matters is that the cat catches mice'.

Two- and three-value boundary analysis

With BVA, we think of the boundary as a dividing line between two things. Hence, we have a value on each side of the boundary, but the boundary itself is not a value:

Invalid	Valid	Invalid
0	1 99	100

Looking at the values for our printer example, 0 is in an invalid partition, 1 and 99 are in the valid partition and 100 is in the other invalid partition. So, the boundary is between the values of 0 and 1, and between the values of 99 and 100. There is a school of thought that regards an actual value as a boundary value. By tradition, these are the values in the valid partition (that is, the values specified). This approach then requires three values for every boundary, so you would have 0, 1 and 2 for the left boundary, and 98, 99 and 100 for the right boundary in this example. The boundary values are said to be on and either side of the boundary and the value that is 'on' the boundary is generally taken to be in the valid partition.

Note that Beizer [1990] and Kaner, Padmanabhan and Hoffman [2013] talk about domain testing, a generalization of EP, with three-value boundaries. Beizer makes a distinction between open and closed boundaries, where a closed boundary is one where the point is included in the domain. So, the convention is for the valid partition to have closed boundaries. You may be pleased to know that you do not have to know this for the exam! Three-value boundary testing is covered in ISO/IEC/IEEE 29119-4 [2021].

So which approach is best? If you use the two-value approach together with EP, you are equally effective and slightly more efficient than the three-value approach. (We will not go into the details here, but this can be demonstrated.) However, three-value BVA may find a defect that would not be found by two-value, so is typically considered more rigorous. In this book we will usually use the two-value approach. In the exam, you may have a question based on either the two-value or the three-value approach, but it should be clear what the correct choice is in either case.

Designing test cases using EP and BVA

Having identified the test conditions that you wish to test, in this instance by using EP and BVA, the next step is to design the test cases. The more test conditions that can be covered in a single test case, the fewer test cases will be needed in order to cover all the conditions. This is usually the best approach to take for positive tests and for tests that you are reasonably confident will pass. However, if a test fails, then we need to find out why it failed. Which test condition was handled incorrectly? We need to get a good balance between covering too many and too few test conditions in our tests.

Let's look at how one test case can cover one or more test conditions. Using the bank balance example, our first test could be of a new customer with a balance of $500. This would cover a balance in the partition from $100.01 to $999.99 and an output partition of a 5% interest rate. We would also be covering other partitions that we have not discussed yet, for example a valid customer, a new customer, a customer with only one account, etc. All of the partitions covered in this test are valid partitions.

When we come to test invalid partitions, the safest option is probably to try to cover only one invalid test condition per test case. This is because programs may stop processing input as soon as they encounter the first problem. So, if you have an invalid customer name, invalid address and invalid balance, you may get an error message saying 'Invalid input' and you do not know whether the test has detected only one invalid input or all of them. This is also why specific error messages are much better than general ones!

However, if it is known that the software under test is required to process all input regardless of its validity, then it is sensible to continue as before and design test cases that cover as many invalid conditions in one go as possible. For example, if every invalid field in a form has some red text above or below the field saying that this field is invalid and why, then you know that each field has been checked, so you have tested all the error processing in one test case. In either case, there should be separate test cases covering valid and invalid conditions.

To cover the boundary test cases, it may be possible to combine all of the minimum valid boundaries for a group of fields into one test case and also the maximum valid boundary values. The invalid boundaries could be tested together if the validation is done on every field; otherwise, they should be tested separately, as with the invalid partitions.

Why do both equivalence partitioning and boundary value analysis?

Technically, because every boundary is in some partition, if you did only BVA you would also have tested every equivalence partition. However, this approach may cause problems if that value fails—was it only the boundary value that failed or did the whole partition fail? Also, by testing only boundaries we would probably not give the users much confidence as we are using extreme values rather than normal values. The boundaries may be more difficult (and therefore more costly) to set up as well.

For example, in the printer copies example described earlier we identified the following boundary values:

Invalid	Valid	Invalid
0	1 99	100

Suppose we test only the valid boundary values 1 and 99 and nothing in between. If both tests pass, this seems to indicate that all the values in between should also work. However, suppose that one page prints correctly, but 99 pages do not. Now we do not know whether any set of more than one page works, so the first thing we would do would be to test for say 10 pages, which is a value from the equivalence partition.

We recommend that you test the partitions separately from boundaries. This means choosing partition values that are NOT boundary values.

However, if you use the three-value boundary value approach, then you would have valid boundary values of 1, 2, 98 and 99, so having a separate equivalence value in addition to the extra two boundary values would not give much additional benefit. But notice that one equivalence value, for example 10, replaces both of the extra two boundary values (2 and 98). This is why EP with two-value BVA is more efficient than three-value BVA.

Which partitions and boundaries you decide to exercise (you do not need to test them all), and which ones you decide to test first, depend on your test objectives. If your goal is the most thorough approach, then follow the procedure of testing valid partitions first, then invalid partitions, then valid boundaries and finally invalid boundaries. However, if you are under time pressure (and who isn't?), then you can't test as much as you would like. Now your test objectives will help you decide what to test. If you are after user confidence of typical transactions with a minimum number of tests, you might do valid partitions only. If you want to find as many defects as possible as quickly as possible, you may start with boundary values, both valid and invalid. If you want confidence that the system will handle bad inputs correctly, you may do mainly invalid partitions and boundaries. Your previous experience of types of defects can help you find similar defects, for example if there are typically a lot of boundary defects, then you would start by testing boundaries.

EP and BVA are described in most testing books, including Myers *et al.* [2011], Copeland [2004], Craig [2002], Kaner *et al.* [1999, 2013], Koomen *et al.* [2006], O'Regan [2019] and Jorgensen [2014]. EP and BVA are described in ISO/IEC/IEEE 29119-4 [2021], including designing tests and measuring coverage.

4.2.3 Decision table testing

The techniques of EP and BVA are often applied to specific situations or inputs. However, if different combinations of inputs result in different actions being taken, this can be more difficult to show using EP and BVA, which tend to be more focused on the user interface. The other two specification-based techniques, **decision table testing** and state transition testing, are more focused on business logic or business rules.

A decision table is a good way to deal with combinations of things (for example, inputs). This technique is sometimes also referred to as a 'cause–effect' table. The reason for this is that there is an associated logic diagramming technique called 'cause–effect graphing' which was sometimes used to help derive the decision table. Myers [2011] describes this as a combinatorial logic network. However, most people find it more useful just to use the table described in Copeland [2004].

> **Decision table testing** A black-box test technique in which test cases are designed to exercise the combinations of conditions and the resulting actions shown in a decision table.

If you begin using decision tables to explore what the business rules are that should be tested, you may find that the analysts and developers find the tables very helpful and want to begin using them too. Do encourage this, as it will make your job easier in the future. Decision tables provide a systematic way of stating complex business rules, which is useful for developers as well as for testers. Decision tables can be used in test design whether or not they are used in development, as they help testers explore the effects of combinations of different inputs and other software states that must correctly implement business rules. Helping the developers do a better job can also lead to better relationships with them.

Testing combinations can be a challenge, as the number of combinations can often be huge. Testing all combinations may be impractical, if not impossible. We have to be satisfied with testing just a small subset of combinations, but it's not easy to choose which combinations to test and which not to test. If you don't have a systematic way of selecting combinations, an arbitrary subset will be used, and this may well result in an ineffective test effort.

Decision tables aid the systematic selection of effective test cases and can have the beneficial side-effect of finding problems and ambiguities in the specification. It is a technique that works well in conjunction with EP. The combination of conditions explored may be combinations of equivalence partitions.

In addition to decision tables, there are other techniques that deal with testing combinations of things: pairwise testing and orthogonal arrays. These are described in Copeland [2004]. Other sources of techniques are Pol *et al.* [2001] and Black [2007]. Decision tables and cause–effect graphing are described in ISO/IEC/IEEE 29119-4 [2021], including designing tests and measuring coverage.

Using decision tables for test design

The first task is to identify a suitable function or subsystem that has a behaviour which reacts according to a combination of inputs or events. The behaviour of interest must not be too extensive (that is, should not contain too many inputs) otherwise the number of combinations will become cumbersome and difficult to manage. It is better to deal with large numbers of conditions by dividing them into subsets and dealing with the subsets one at a time. A risk-based approach may be used to select which are the most important combinations to test, or a minimized table with only selected (most important) conditions may be used.

Once you have identified the aspects that need to be combined, then you put them into a table, listing all the combinations of True and False for each of the aspects. Take an example of a loan application, where you can enter the amount of the monthly repayment or the number of years you want to take to pay it back (the term of the loan). If you enter both, we assume that the system will make a compromise between the two if they conflict. The two conditions are the repayment amount and the term, so we put them in a table (refer to Table 4.2).

TABLE 4.2 Empty decision table

Conditions	Rule 1	Rule 2	Rule 3	Rule 4
Repayment amount has been entered				
Term of loan has been entered				

Next, we will identify all of the combinations of True and False (refer to Table 4.3). With two conditions, each of which can be True or False, we will have four combinations (two to the power of the number of things to be combined). Note that if we have three things to combine, we will have eight combinations; with four things, we will have 16, etc. This is why it is good to tackle small sets of combinations at a time. In order to keep track of which combinations we have, we will alternate True and False on the bottom row, put two Trues and then two Falses on the row above the bottom row, etc., so the top row will have all Trues and then all Falses (and this principle applies to all such tables).

TABLE 4.3 Decision table with input combinations

Conditions	Rule 1	Rule 2	Rule 3	Rule 4
Repayment amount has been entered	T	T	F	F
Term of loan has been entered	T	F	T	F

We are using T and F for the inputs in our example; we could also have used Y and N or 1 (the number one) and 0 (zero) respectively. The conditions can also be numbers, numeric ranges or enumerated types (for example, red, green or blue). If Boolean values (that is, True and False) are used, it is called a limited entry decision table; extended entry tables allow other values.

The next step (at least for this example) is to identify the correct outcome for each combination (refer to Table 4.4). In this example, we can enter one or both of the two fields. Each combination is sometimes referred to as a rule. In this example, we are using Y for the actions which should occur and leaving it blank if that action should not occur. Other options are to use X or 1 if an action should occur, and N, F, '–' or 0 for actions that should not occur. We could use a number or range of numbers if an outcome occurs for those numbers and does not occur for others. You could also use discrete values, for example an action should occur if an input is red but does not occur if it is green or yellow.

TABLE 4.4 Decision table with combinations and outcomes

Conditions	Rule 1	Rule 2	Rule 3	Rule 4
Repayment amount has been entered	T	T	F	F
Term of loan has been entered	T	F	T	F
Actions/Outcomes				
Process loan amount	Y	Y		
Process term	Y		Y	

At this point, we may now realize that we had not thought about what happens if the customer doesn't enter anything in either of the two fields. The table has highlighted a combination that was not mentioned in the specification for this example. We could assume that this combination should result in an error message, so we need to add another action (refer to Table 4.5). This highlights the strength of this technique to discover omissions and ambiguities in specifications. It is not unusual for some combinations to be omitted from specifications; therefore, this is also a valuable technique to use when reviewing the test basis.

TABLE 4.5 Decision table with additional outcome

Conditions	Rule 1	Rule 2	Rule 3	Rule 4
Repayment amount has been entered	T	T	F	F
Term of loan has been entered	T	F	T	F
Actions/Outcomes				
Process loan amount	Y	Y		
Process term	Y		Y	
Error message				Y

Suppose we change our example slightly, so that the customer is not allowed to enter both repayment and term. Now our table will change, because there should also be an error message if both are entered, so it will look like Table 4.6.

TABLE 4.6   Decision table with changed outcomes

Conditions	Rule 1	Rule 2	Rule 3	Rule 4
Repayment amount has been entered	T	T	F	F
Term of loan has been entered	T	F	T	F
Actions/Outcomes				
Process loan amount		Y		
Process term			Y	
Error message	Y			Y

You might notice now that there is only one Yes in each column, that is, our actions are mutually exclusive: only one action occurs for each combination of conditions. We could represent this in a different way by listing the actions in the cell of one row, as shown in Table 4.7. Note that if more than one action results from any of the combinations, then it would be better to show them as separate rows rather than combining them into one row.

TABLE 4.7   Decision table with outcomes in one row

Conditions	Rule 1	Rule 2	Rule 3	Rule 4
Repayment amount has been entered	T	T	F	F
Term of loan has been entered	T	F	T	F
Actions/Outcomes				
Result	Error message	Process loan amount	Process term	Error message

The final step of this technique is to write test cases to exercise each of the four rules in our table.

In this example we started by identifying the input conditions and then identifying the outcomes. However, in practice it might work the other way around—we can see that there are a number of different outcomes and have to work back to understand what combination of input conditions actually drive those outcomes. The technique works just as well doing it in this way and may well be an iterative approach as you discover more about the rules that drive the system.

Credit card worked example

Let's look at another example. If you are a new customer opening a credit card account, you will get a 15% discount on all your purchases today. If you are an existing customer and you hold a loyalty card, you get a 10% discount. If you have a coupon, you can get 20% off today (but it cannot be used with the new customer discount). Discount amounts are added, if applicable. This is shown in Table 4.8.

TABLE 4.8   Decision table for credit card example

Conditions	Rule 1	Rule 2	Rule 3	Rule 4	Rule 5	Rule 6	Rule 7	Rule 8
New customer (15%)	T	T	T	T	F	F	F	F
Loyalty card (10%)	T	T	F	F	T	T	F	F
Coupon (20%)	T	F	T	F	T	F	T	F
Actions								
Discount (%)	X	X	20	15	30	10	20	0

In Table 4.8, the conditions and actions are listed in the left-hand column. All the other columns in the decision table each represent a separate rule, one for each combination of conditions. We may choose to test each rule/combination and if there are only a few, this will usually be the case. However, if the number of rules/combinations is large we are more likely to sample them by selecting a rich subset for testing.

Note that we have put X for the discount for two of the columns (Rules 1 and 2). This means that this combination should not occur. You cannot be both a new customer and already hold a loyalty card! There should be an error message stating this, but even if we don't know what that message should be, it will still make a good test.

Because we have exactly the same action (and presumably the same error message) for both columns 1 and 2, we can see that having a coupon makes no difference to the outcome, that is, we do not care whether you have a coupon or not, as it makes no difference to the actions or outcome. This means that we can slightly collapse or shorten the table as shown in Table 4.9. We have put a dash (–) to show that it does not matter whether coupon is True or False. This is not very significant for such a small example. We could also put 'N/A' (Not Applicable) to show that this value does not affect the outcome. We can combine two or more columns in this way, as long as they all have the same resulting actions or outcomes. Refer to Copeland [2004] for more information.

TABLE 4.9   Collapsed decision table for credit card example

Conditions	Rule 1	Rule 3	Rule 4	Rule 5	Rule 6	Rule 7	Rule 8
New customer (15%)	T	T	T	F	F	F	F
Loyalty card (10%)	T	F	F	T	T	F	F
Coupon (20%)	–	T	F	T	F	T	F
Actions							
Discount (%)	X	20	15	30	10	20	0

With more complex decision tables, being able to collapse the table by combining columns (or deleting a column) can make the table much easier to read and understand. For clarity, we have just removed 'Rule 2' and kept the other rule numbers as they were. This could be changed to make the rule numbers sequential in the final table. More information about collapsing decision tables is covered in the ISTQB Advanced Level Test Analyst Syllabus.

We have made an assumption in Rule 3. Since the coupon has a greater discount than the new customer discount, we assume that the customer will choose 20% rather than 15%. We cannot add them, since the coupon cannot be used with the new customer discount. The 20% action is an assumption on our part, and we should check that this assumption (and any other assumptions that we make) is correct, by asking the person who wrote the specification or the users.

For Rule 5, however, we can add the discounts, since both the coupon and the loyalty card discount should apply (at least that is our assumption).

Rules 4, 6 and 7 have only one type of discount and Rule 8 has no discount, so is 0%.

If we are applying this technique thoroughly, we would have one test for each column or rule of our decision table. The advantage of doing this is that we may test a combination of things that otherwise we might not have tested and that could find a defect. Coverage of decision table testing is measured by the number of columns that have at least one test case, divided by the total number of columns.

However, if we have a lot of combinations, it may not be possible or sensible to test every combination. Or if we are time-constrained, we may not have time to test all combinations. Do not just assume that all combinations need to be tested; it is better to prioritize and test the most important combinations. Having the full table enables us to see which combinations we decided to test and which not to test this time.

There may also be many different actions as a result of the combinations of conditions. Here we just had one: the discount to be applied. The decision table shows which actions apply to each combination of conditions.

In the example above all the conditions are binary, that is, they have only two possible values: True or False (or, if you prefer Yes or No). Often it is the case that conditions are more complex, having potentially many possible values. Where this is the case the number of combinations is likely to be very large, so the combinations may only be sampled rather than exercising all of them.

4.2.4 State transition testing

State transition testing A black-box test technique in which test cases are designed to exercise elements of a state transition model. (Also known as finite state testing)

State transition testing is used where some aspect of the system can be described in what is called a 'finite state machine'. This simply means that the system can be in a limited (finite) number of different states, and the transitions from one state to another are determined by the rules of the 'machine'. This is the model on which the system and the tests are based. Any system where you get a different output for the same input, depending on what has happened before, is a finite state system. A finite state system is often shown as a state diagram (refer to Figure 4.2).

For example, if you request to withdraw $100 from a bank ATM, you may be given cash. Later you may make exactly the same request but be refused the money (because your balance is insufficient). This later refusal is because the state of your bank account has changed from having sufficient funds to cover the withdrawal to having insufficient funds. When the same event can result in two different transitions, depending on some True/False condition (in this case sufficient funds), this is referred to as a 'guard condition' on the transition. The funds will be dispensed only when the guard condition of sufficient funds is True. Guard conditions are shown on a state diagram in square brackets by the transition that they are guarding. The transaction that caused your account to change its state was probably the earlier withdrawal. A state diagram can represent a model from the point of view of the system, the account or the customer.

Another example is a word processor. If a document is open, you are able to close it. If no document is open, then Close is not available. After you choose Close once, you cannot choose it again for the same document unless you open that document. A document thus has two states: open and closed.

A state transition model has four basic parts:

- The states that the software may occupy (open/closed or funded/insufficient funds).
- The transitions from one state to another (not all transitions are allowed).
- The events that cause a transition (closing a file or withdrawing money).
- The actions that result from a transition (an error message or being given your cash).

Note that in any given state, one event can cause only one action, but that the same event from a different state may cause a different action and a different end state.

We will look first at test cases that execute valid state transitions.

Figure 4.2 shows an example of entering a PIN to access a bank account. The states are shown as circles, the transitions as lines with arrows and the events as the text near the transitions. We have not shown the actions explicitly on this diagram, but they would be a message to the customer saying things such as 'Please enter your PIN'.

The state diagram shows seven states but only four possible events (Card inserted, Enter PIN, PIN OK and PIN not OK). We have not specified all of the possible transitions here: there would also be a time-out from 'wait for PIN' and from the three tries. The system would go back to the start state after the time had elapsed and would probably eject the card. There would also be a transition from the 'eat card' state back to the start state. We have not specified all the possible events either—there would be a 'cancel' option from 'wait for PIN' and from the three tries, which would also go back to the start state and eject the card. The 'access account' state would be the beginning of another state diagram showing the valid transactions that could now be performed on the account. This state diagram, even though it is incomplete, still gives us information on which to design some useful tests and to explain the state transition technique.

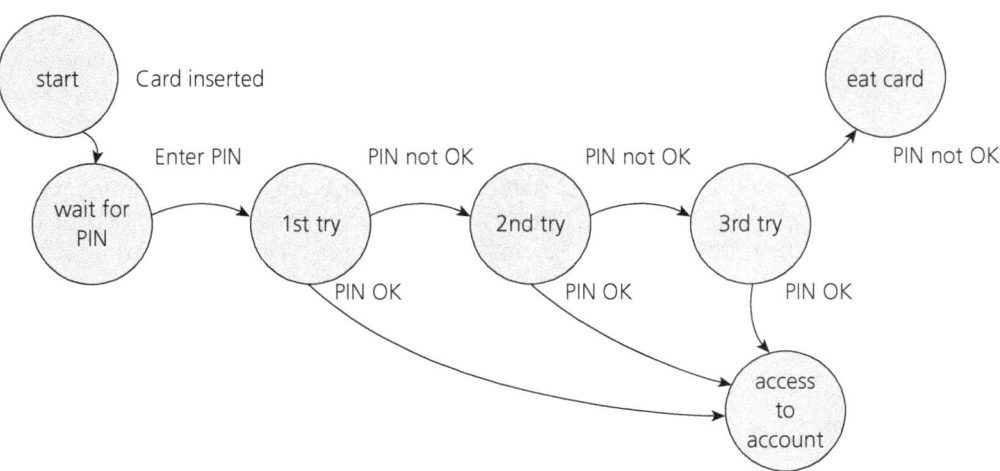

FIGURE 4.2   State diagram for PIN entry

In deriving test cases, we may start with a typical scenario. A sensible first test case here would be the normal situation, where the correct PIN is entered the first time. To be more thorough, we may want to make sure that we cover every state (that is, at least one test goes through each state) or we may want to cover every transition. A second test (to visit every state) would be to enter an incorrect PIN each time, so that the system eats the card. We still have not tested every transition yet. In order to do that, we would want a test where the PIN was incorrect the first time but OK the second time, and another test where the PIN was correct on the third try. These tests are probably less important than the first two.

Note that a transition does not need to change to a different state (although all of the transitions shown in the figure do go to a different state). There could be a transition from 'access to account' which just goes back to 'access to account' for an action such as 'request balance'.

Test conditions can be derived from the state diagram in various ways. Each state can be noted as a test condition, as can each transition. In the Syllabus, we need to be able to identify the coverage of a set of tests in terms of states or transitions.

Tests can be designed to cover a typical sequence of states, to visit all states, to exercise every transition, or to test invalid transitions (we look at coverage criteria for state transition testing shortly).

One of the advantages of the state transition technique is that the model can be as detailed or as abstract as you need it to be. Where a part of the system is more important (that is, requires more testing) a greater depth of detail can be modelled. Where the system is less important (requires less testing), the model can use a single state to signify what would otherwise be a series of different states.

Testing for invalid transitions

Deriving tests only from a state diagram or chart (also known as a state graph) is very good for seeing the valid transitions, but we may not easily see the negative tests, where we try to generate invalid transitions. To see the total number of combinations of states and transitions, both valid and invalid, a state table is useful.

The state table lists all the states down one side of the table and all the events that cause transitions along the top (or vice versa). Each cell then represents a state-event pair. The content of each cell indicates which state the system will move to when the corresponding event occurs while in the associated state. This will include possible erroneous events, which are events that are not expected to happen in certain states. These are negative test conditions.

Table 4.10 lists the states in the first column and the possible events across the top row. So, for example, if the system is in State 1, inserting a card will take it to State 2. If we are in State 2, and a valid PIN is entered, we go to State 6 to access the account. In State 2 if we enter an invalid PIN, we go to State 3. We have put a dash in the cells that should be impossible, that is, they represent invalid transitions from that state.

TABLE 4.10 State table for the PIN example

	Insert card	**Valid PIN**	**Invalid PIN**
S1) Start state	S2	–	–
S2) Wait for PIN	–	S6	S3
S3) 1st try invalid	–	S6	S4
S4) 2nd try invalid	–	S6	S5
S5) 3rd try invalid	–	–	S7
S6) Access account	–	?	?
S7) Eat card	S1 (for new card)	–	–

We have put a question mark for two cells, where we enter either a valid or invalid PIN when we are accessing the account. Perhaps the system will take our PIN number as the amount of cash to withdraw? It might be a good test! Most of the other invalid cells would be physically impossible in this example. Invalid (negative) tests will attempt to generate invalid transitions, transitions that should not be possible (but often make good tests when it turns out they are possible).

State transition testing is used for screen-based applications, for example an ATM, and also within the embedded software industry. It is useful for modelling business scenarios which have specific states, or for testing navigation around screens.

A more extensive description of state machines is found in Marick [1994]. State transition testing is also described in Craig and Jaskiel [2002], Copeland [2004], Beizer [1990], Broekman and Notenboom [2003] and Black [2007]. State transition testing is described in ISO/IEC/IEEE 29119-4 [2015], including designing tests and coverage measures.

Coverage criteria for state transition testing

The Syllabus lists three coverage criteria: all states, valid transitions and all transitions, and they are exactly what they say they are.

All state coverage is the number of states (in a state transition diagram or state table) that have been exercised by at least one test, compared to the total number of states (i.e. the percentage of states that have been tested). To achieve 100% all state coverage, each state must be visited by at least one test. This is the weakest of these three coverage criteria.

Valid transition coverage is the number of valid transitions that have been exercised, compared to the total number of valid transitions in the model we are considering, also expressed as a percentage. To achieve 100% valid transition coverage, each valid transition must be exercised by at least one test. Note that there may be some states that have more than one valid transition to another state; we can achieve 100% all state coverage without exercising every valid transition. This is why valid transition coverage is stronger than all state coverage: if you cover every valid transition, then you will also have covered every state.

All transitions coverage includes not just valid transitions but also invalid transitions, which are shown in the state table. The coverage is defined as the number of all transitions exercised (or attempted—some invalid transitions are physically impossible), compared to the total number of transitions, as a percentage. Some invalid transitions will generate an error message; in this case we can design a test to exercise this transition. As with other negative tests, it is a good idea to try to test only one invalid transition per test, otherwise defects may be hiding behind a defect found by the first invalid transition; this is called defect masking.

Probably the most commonly used state transition coverage criteria is valid transitions coverage, also known as 0-switch coverage. For more critical systems, a stronger coverage criterion may be needed.

Going beyond the level expected in the Syllabus, we can also consider transition pairs and triples and so on. Coverage of all individual valid transitions is 0-switch coverage, coverage of transition pairs is 1-switch coverage, coverage of transition triples is 2-switch coverage, etc. Measuring how much you have tested (covered) is getting close to a white-box perspective. However, state transition testing is regarded as a black-box technique. More information on coverage criteria for state transition testing is covered in the ISTQB Advanced Level Test Analyst Syllabus.

4.3 WHITE-BOX TEST TECHNIQUES

> **SYLLABUS LEARNING OBJECTIVES FOR 4.3 WHITE-BOX TEST TECHNIQUES (K2)**
>
> FL-4.3.1   **Explain statement testing (K2)**
>
> FL-4.3.2   **Explain branch testing (K2)**
>
> FL-4.3.3   **Explain the value of white-box testing (K2)**

In this section we will look in detail at the concept of coverage and how it can be used to measure some aspects of the thoroughness of testing. In order to see how coverage actually works, we will use some code-level examples (although coverage also applies to other levels such as business procedures). In particular, we will show how to measure coverage of statements and branches, and how to write test cases to extend coverage if it is not 100%.

As mentioned, we will illustrate white-box test techniques that would typically apply at the component level of testing. At this level, white-box techniques focus on the structure of a software component, such as statements, decisions, branches or even distinct paths. These techniques can also be applied at the integration level. At this level, white-box techniques can examine structures such as a call tree, which is a diagram that shows how modules call other modules. These techniques can also be applied at the system level. At this level, white-box techniques can examine structures such as a menu structure, business process or web page structure.

There are also more advanced white-box test techniques at component level, particularly for safety-critical, mission-critical or high-integrity environments to achieve higher levels of coverage. For more information on these techniques refer to the ISTQB Advanced Level Technical Test Analyst Syllabus.

In this section, look for the definitions of the Glossary terms **branch coverage** and **statement coverage**.

White-box test techniques serve two purposes: coverage measurement and structural test case design. They are often used first to assess the amount of testing performed by tests derived from black-box test techniques, that is, to assess coverage. They are then used to design additional tests with the aim of increasing the coverage.

White-box test techniques used to design tests are a good way of generating additional test cases that are different from existing tests. They can help ensure more breadth of testing, in the sense that test cases that achieve 100% coverage in any measure will be exercising all parts of the software from the point of view of the items being covered.

What is coverage?

Coverage measures the amount of testing performed by a set of tests that may have been derived in a different way, for example using black-box techniques. Wherever we can count things and can tell whether or not each of those things has been tested by some test, then we can measure coverage. The basic coverage measure is:

$$\text{Coverage} = \frac{\text{Number of coverage items exercised}}{\text{Total number of coverage items}} \times 100\%$$

where the coverage item is whatever we have been able to count and see whether a test has exercised or used this item.

A Christmas gift to one of the authors was a map of the world where you can scratch off the countries you have visited. This is a very good analogy for coverage. If I visit a small country, for example Belgium, it seems quite OK to scratch off the whole country. But if I visit the US, Canada, Australia or China then do I scratch off the whole country? Even if I have only visited one or a few cities, I can still count the whole country, and this seems not quite right. But this is 'country coverage'. I can count the countries I have visited, divided by the total number of countries in the world, and get a percentage, for example around 18% for visiting 36 countries. However, if you look at the map, it looks like I have covered around 25% of the area (since I have been to the US, Canada, Australia, Brazil and China). If I were to do area-of-the-map coverage, I would get a higher coverage percentage, but I would not have seen any more of the world. Now that I have visited Greenland, I increase my country coverage by less than 1%, but increase my area coverage by nearly 2%. Why are we talking about this map? Because it highlights some of the same aspects and pitfalls of coverage for components and systems.

There are several dangers (pitfalls or caveats) in using coverage measures:

- 100% coverage does *not* mean 100% tested! Coverage techniques measure only one dimension of a multi-dimensional concept. Country coverage is a higher level than state, county or city coverage.

- Two different test cases may achieve exactly the same coverage but the input data of one may find an error that the input data of the other does not. If you do not execute a line or block of code that contains a bug, you are guaranteed not to see the failures that bug can cause. However, just because you did execute that line or block does not guarantee that you will see the failures that bug can cause. There may be many different data combinations that exercise the same True/False decision outcome, but some will cause a failure and others will not.

- Coverage looks only at what *has* been written, that is the code itself. It cannot say anything about the software that has *not* been written. If a specified function has not been implemented, black-box test techniques will reveal this. If a function was omitted from the specification, then experience-based techniques may find it. But white-box techniques can only look at a structure which is already there.

- Just because some coverage item has been covered, this does NOT mean that this part of the system is actually doing what it should. Coverage only assesses whether or not you have exercised something, not whether that test passed or failed or whether it was a good test worth running. Coverage says nothing about the quality of either the system or the tests.

So, is coverage worth measuring? Yes, coverage can be useful. It is a way of assessing one aspect of thoroughness. But it is best used when you understand exactly what you are measuring and are aware of the pitfalls, particularly if you are reporting coverage to stakeholders.

Types of coverage

Coverage can be measured based on a number of different structural elements in a system or component. Coverage can be measured at component testing level, integration testing level or at system or acceptance testing levels. For example, at system or acceptance level, the coverage items may be requirements, menu options, screens or typical business transactions. Other coverage measures include things such as database structural elements (records, fields and sub-fields) and files or neuron coverage of neural networks. It is worth checking for any new tools, as the test tool market develops quite rapidly.

At integration level, we could measure coverage of interfaces or specific interactions that have been tested. The call coverage of APIs, modules, objects or procedure calls can also be measured (and is supported by tools to some extent).

There is good tool support for code coverage, that is, for component-level testing. We can measure coverage for each of the black-box test techniques as well:

- Equivalence partitioning: percentage of equivalence partitions exercised (we could measure valid and invalid partition coverage separately if this makes sense).

- Boundary value analysis: percentage of boundaries exercised (we could also separate valid and invalid boundaries if we wished).

- Decision tables: percentage of business rules or decision table columns tested.

- State transition testing: there are a number of possible coverage measures:
  - percentage of states visited
  - percentage of valid transitions exercised (this is known as 0-switch coverage)
  - percentage of all transitions exercised or attempted
  - percentage of pairs of valid transitions exercised ('transition pairs' or 1-switch coverage), and longer series of transitions, such as transition triples, quadruples, etc.
  - percentage of invalid transitions exercised (from the state table).

The coverage measures for black-box techniques would apply at whichever test level the technique has been used (for example, system or component level).

When coverage is discussed by product owners, business analysts, system testers or users, it most likely refers to the percentage of requirements that have been tested by a set of tests. This may be measured by a tool such as a requirements management tool or a test management tool.

However, when coverage is discussed by programmers, it most likely refers to the coverage of code, where the structural elements can be identified using a tool. We will cover statement and branch coverage shortly. However, at this point, note that the word coverage is often misused to mean 'How many or what percentage of tests have been run?' This is **NOT** what the term coverage refers to. Coverage is the coverage **of** something else **by** the tests. The percentage of tests run should be called test completeness or something similar.

Statements and branches (the control flow path taken as the result of a decision outcome) are both structures that can be measured in code and there is good tool support for these coverage measures. Code coverage is normally done in component and component integration testing, if it is done at all. If someone claims to have achieved code coverage, it is important to establish exactly what elements of the code have been covered, as statement coverage (often what is meant) is significantly weaker than branch coverage or some of the other code coverage measures.

How to measure coverage

For most practical purposes, coverage measurement is something that requires tool support. However, knowledge of the steps typically taken to measure coverage is useful in understanding the relative merits of each technique. Our example assumes an intrusive coverage measurement tool that alters the code by inserting instrumentation:

1 Decide on the structural element to be used, that is, the coverage items to be counted.

2 Count the structural elements or items.

3 Instrument the code.

4 Run the tests for which coverage measurement is required.

5 Using the output from the instrumentation, determine the percentage of elements or items exercised.

Instrumenting the code (step 3) involves inserting code alongside each structural element in order to record when that structural element has been exercised. Determining the actual coverage measure (step 5) is then a matter of analyzing the recorded information.

Coverage measurement of code is best done using tools and there are a number of such tools on the market. These tools can help to support increased quality and productivity of testing. They support increased quality by ensuring that more structural aspects are tested, so defects on those structural paths can be found (and fixed, otherwise quality has not increased). They support increased productivity and efficiency by highlighting tests that may be redundant, that is, testing the same structure as other tests, although this is not necessarily a bad thing, since we may find a defect testing the same structure with different data.

In common with all white-box test techniques, code coverage techniques are best used on areas of software code where more thorough testing is required. Safety-critical code, code that is vital to the correct operation of a system, and complex pieces of code are all examples of where white-box techniques are particularly worth applying. For example, some standards for safety-critical systems, such as avionics and vehicle control, require white-box coverage for certain types of system. White-box test techniques should normally be used in addition to black-box and experience-based test techniques rather than as an alternative to them.

White-box test design

If you are aiming for a given level of coverage (say 95%) but you have not reached your target (for example, you only have 87% so far), then additional test cases can be designed with the aim of exercising some or all of the structural elements not yet reached. This is white-box or structure-based test design. These new tests are then run through the instrumented code and a new coverage measure is calculated. This is repeated until the required coverage measure is achieved (or until you decide that your goal was too ambitious!). Ideally all the tests ought to be run again on the un-instrumented code.

We will look at some examples of structure-based coverage and test design for statement and branch testing shortly.

4.3.1 Statement testing and coverage

Statement coverage is calculated by:

$$\text{Statement coverage} = \frac{\text{Number of statements exercised}}{\text{Total number of statements}} \times 100\%$$

Studies and experience in the industry have indicated that what is considered reasonably thorough black-box testing may actually achieve only 60% to 75% statement coverage. Typical ad hoc testing is likely to achieve only around 30%, leaving 70% of the statements untested. Executable statements are converted to running code by the compiler, and exclude comments, data declarations, etc.

Different coverage tools may work in slightly different ways, so they may give different coverage figures for the same set of tests on the same code, although at 100% coverage they should be the same.

We will illustrate the principles of coverage on code. In order to explain our examples, we will use two types of code examples, one a basic pseudo-code—this is not any specific programming language, but should be readable and understandable to you, even if you have not done any programming yourself—and the second is more like JavaScript. Both give the same control flow. We have omitted the set-up code that is needed to actually run the code, to concentrate on the logic.

For example, consider Code samples 4.1a and 4.1b.

```
READ A
READ B
IF A > B THEN C = 0
ENDIF
```

Code sample 4.1a

```
let a = Number(args[2])
let b = Number(args[3])
if (a > b) {
        c = 0
}
```

Code sample 4.1b

To achieve 100% statement coverage of this code segment just one test case is required, one which ensures that variable A contains a value that is greater than the value of variable B, for example A = 12 and B = 10. Note that here we are doing structural test *design* first, since we are choosing our input values in order to ensure statement coverage.

Let's look at an example where we measure coverage first. In order to simplify the example, we will regard each line as a statement. Different tools and methods may count different things as statements, but the basic principle is the same however they are counted. A statement may be on a single line, or it may be spread over several lines. One line may contain more than one statement, just one statement or only part of a statement. Some statements can contain other statements inside them. In Code samples 4.2, we have two read statements, one assignment statement and then one IF statement on three lines, but the IF statement contains another statement (print) as part of it.

```
1 READ A
2 READ B
3 C = A + 2*B
4 IF C > 50 THEN
5        PRINT 'Large C'
6 ENDIF
```

Code sample 4.2a

```
1 let a = Number(args[2])
2 let b = Number(args[3])
3 let c = a + 2*b
4 if (c > 50) {
5     console.log('C large')
6 }
```

Code sample 4.2b

Although it is not completely correct, we have numbered each line and will regard each line as a statement. Some tools may group statements that would always be executed together in a basic block which is regarded as a single statement. However, we will just use numbered lines to illustrate the principle of coverage of statements (lines). Let's analyze the coverage of a set of tests on our six-statement program:

TEST SET 1

Test 1_1: A = 2; B = 3
Test 1_2: A = 0; B = 25
Test 1_3: A = 47, B = 1

Which statements have we covered?

- In Test 1_1, the value of C will be 8, so we will cover the statements on lines 1 to 4 and line 6.
- In Test 1_2, the value of C will be 50, so we will cover exactly the same statements as Test 1_1.
- In Test 1_3, the value of C will be 49, so again we will cover the same statements.

Since we have covered five out of six statements, we have 83% statement coverage (with three tests). What test would we need in order to cover statement 5, the one statement that we haven't exercised yet? How about this one:

Test 1_4: A = 20; B = 25

This time the value of C is 70, so we will print 'Large C' and we will have exercised all six of the statements, so now statement coverage = 100%. Notice that we measured coverage first, and then designed a test to cover the statement that we had not yet covered.

Note that Test 1_4 on its own is more effective (towards our goal of achieving 100% statement coverage) than the first three tests together. Just taking Test 1_4 on its own is also more efficient than the set of four tests, since it has used only one test instead of four. Being more effective and more efficient is the mark of a good test technique.

4.3.2 Branch testing and coverage

A branch is a transfer of control between two nodes of a control flow graph—we will show an example of a control flow graph shortly. This type of graph shows how control flows from one statement to another within a software program. A simple statement, for example a Print statement, is executed after the one before it, and the statement following it is executed next. A decision statement is a different type of statement, as the statement coming after it depends on something (the outcome of that decision). Examples include an IF statement, a loop control statement (for example, DO–WHILE or REPEAT–UNTIL), or a CASE statement, where there are two or more possible exits or outcomes from the statement. With an IF statement, the exit can either be True or False, depending on the value of the logical condition that comes after IF. With a loop control statement, the outcome is either to perform the code within the loop or not—again a True or False exit. A branch is the route taken to the next statement, shown as a line on the control flow graph.

Note that the previous Syllabus used decision coverage rather than branch coverage. These two measures are very closely related, since the outcome of a decision, that is, a True/False decision, results in control taking one of two outcomes; the path that is taken is a branch. Any branch coming from a decision statement is a conditional branch, since the outcome of the decision (based on a logical condition) decides which way it goes. However, there are also unconditional branches, such as the control path after sequential statements going to the next statement. Thus, there will usually be more branches than decision outcomes. The important thing is that at 100% coverage, these two ways of measuring coverage are equivalent. If you have 100% decision coverage, you always have 100% branch coverage, and vice versa.

| **Branch coverage**   The coverage of branches in a control flow graph. |

Branch coverage is calculated by:

$$\text{Branch coverage} = \frac{\text{Number of branches exercised}}{\text{Total number of branches}} \times 100\%$$

What feels like reasonably thorough black-box testing may achieve only 40 to 60% branch coverage. Typical ad hoc testing may cover only 20% of the branches, leaving 80% of the possible branches untested. Even if your testing seems reasonably thorough from a functional or black-box perspective, you may have only covered two-thirds or three-quarters of the branches. Branch coverage is stronger than statement coverage. It 'subsumes' statement coverage, which means that 100% branch coverage always guarantees 100% statement coverage. Any stronger coverage measure may require more test cases to achieve 100% coverage.

Let's go back to Code samples 4.2 again. We saw earlier that just one test case was required to achieve 100% statement coverage. However, branch coverage requires each decision to have had both a True and a False outcome so that the branch from that decision has been exercised. Therefore, to achieve 100% branch coverage, a second test case is necessary where A is less than or equal to B. This will ensure that the decision statement IF A > B has a False outcome, so the False branch has been taken. So, one test is sufficient for 100% statement coverage, but two tests are needed for 100% branch coverage. Note that 100% branch coverage guarantees 100% statement coverage, but *not* the other way around!

Now let us consider slightly different Code samples shown in Code samples 4.3a and 4.3b.

```
1 READ A
2 READ B
3 C = A - 2*B
4 IF C < 0 THEN
5      PRINT 'C negative'
6 ENDIF
```

Code sample 4.3a

```
1 let a = Number(args[2])
2 let b = Number(args[3])
3 let c = a - 2*b
4 if (c < 0) {
5      console.log('C negative')
6 }
```

Code sample 4.3b

Let us suppose that we already have the following test, which gives us 100% statement coverage for Code samples 4.3:

TEST SET 2

Test 2_1: A = 20; B = 15

Which branches have we exercised with our test? The value of C is −10, so the condition C < 0 is True, so we will print 'C negative' and we have exercised the True branch from that decision statement. But we have not exercised the False branch. What other test would we need to exercise the False outcome and to achieve 100% branch coverage?

Before we answer that question, let's have a look at another way to represent this code. Sometimes the structure is easier to see in a control flow diagram (refer to Figure 4.3).

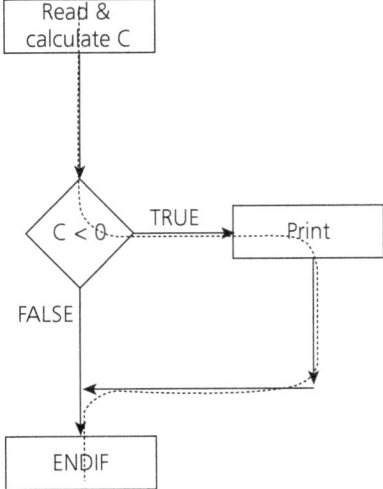

FIGURE 4.3 Control flow diagram for Code samples 4.3

In this diagram we have two conditional branches, one labelled TRUE and the other labelled FALSE. But we also have two unconditional branches: the line from the top box to the decision box, and the line from the Print statement to ENDIF. So there are a total of four branches.

The dotted line shows where Test 2_1 has gone and clearly shows that we have not yet had a test that takes the False exit from the IF statement.

Let's modify our existing test set by adding another test:

TEST SET 2

Test 2_1: A = 20; B = 15
Test 2_2: A = 10; B = 2

This now covers both of the decision outcomes and both conditional branches: True (with Test 2_1) and False (with Test 2_2). We have also covered the unconditional branches. If we were to draw the path taken by Test 2_2, it would be a straight line from the read statement down the False exit and through the ENDIF. Note that we could have chosen other numbers to achieve either the True or False branches.

4.3.3 The value of white-box testing

Coverage is a partial measure of some aspect of the thoroughness of testing. In this section, we have looked at two types of coverage: statement and branch. The value of statement and branch testing is in seeing what new tests are needed in order to achieve a higher level of coverage, whatever dimension of coverage we are looking at. By being more thorough (even in a very limited way), we are leaving less of the component or system completely untested.

One client decided to measure coverage and found that their supposedly thorough tests only reached 60% branch coverage. They decided to write more tests to increase this to 80%. Although they spent three weeks designing and implementing these new tests, they found enough high-severity defects to justify spending that time.

White-box coverage measures and related test techniques are described in ISO/IEC/IEEE 29119-4 [2015]. White-box test techniques are also discussed in Copeland [2003], Myers *et al.* [2011] and Watson *et al.* [1996]. A good description of the graph theory behind structural testing can be found in Jorgensen [2014], and Hetzel [1988] also shows a structural approach. Pol *et al.* [2001] describes a white-box approach called an algorithm test.

4.4 EXPERIENCE-BASED TEST TECHNIQUES

> SYLLABUS LEARNING OBJECTIVES FOR 4.4
> EXPERIENCE-BASED TEST TECHNIQUES (K2)
>
> **FL-4.4.1   Explain error guessing (K2)**
>
> **FL-4.4.2   Explain exploratory testing (K2)**
>
> **FL-4.4.3   Explain checklist-based testing (K2)**

In this section we will look at three experience-based techniques, why and when they are useful, and how they fit with black-box test techniques.

Although it is true that testing should be rigorous, thorough and systematic, this is not all there is to testing. There is a definite role for non-systematic techniques, that is, tests based on a person's knowledge, experience, imagination and intuition. The reason is that some defects are hard to find using more systematic approaches, so a good bug hunter can be very creative at finding those elusive defects.

One aspect which is more difficult for experience-based techniques is the measurement of coverage. In order to measure coverage, we need to have some idea of a full set of things that we could possibly test, coverage then being the percentage of that set that we did test. For example, we could look at the coverage of the items in the test charter, a checklist or a set of heuristics. We should also be able to use the traceability of tests to requirements or user stories and look at coverage of those.

In this section, look for the definitions of the Glossary terms **checklist-based testing**, **error guessing** and **exploratory testing**.

4.4.1 Error guessing

Error guessing is a technique that is good to be used as a complement to other more formal techniques. The success of error guessing is very much dependent on the skill of the tester, as good testers know where the defects are most likely to lurk. Some people seem to be naturally good at testing. Others are good testers because they have a lot of experience either as a tester or working with a particular system and so are able to pin-point its weaknesses. This is why error guessing, used after more formal techniques have been applied to some extent, can be very effective. In using more formal techniques, the tester is likely to gain a better understanding of the system, what it does and how it works. With this better understanding, they are likely to be better at guessing ways in which the system may not work properly.

> **Error guessing** A test technique in which tests are derived on the basis of the tester's knowledge of past failures, or general knowledge of failure modes.

There are no rules for error guessing. The tester is encouraged to think of situations in which the software may not be able to cope. Here are some typical things to try: division by zero, blank (or no) input, empty files and the wrong kind of data (for example, alphabetic characters where numeric are required). If anyone ever says of a system or the environment in which it is to operate 'That could never happen', it might be a good idea to test that condition, as such assumptions about what will and will not happen in the live environment are often the cause of failures.

Error guessing may be based on:

- How the application has worked in the past.
- What types of mistakes the developers tend to make.
- Failures that have occurred in other applications.

The areas where error guessing can be effective include:

- inputs (missing, incorrect format, incorrect number of inputs, correct inputs rejected)
- outputs (missing, wrong value, wrong format)
- computation (wrong formula, incorrect variable used, incorrect operand, e.g. subtract instead of add)

- interfaces (parameter mismatch, e.g. wrong number of parameters, API interface incorrect, communication problems between systems, incompatible types of parameters)

- data (incorrect initialization, wrong type, selected from the wrong file, improperly or incompletely anonymized).

A structured approach to the error-guessing technique is to list possible defects or failures and to design tests that attempt to produce them. These defect and failure lists can be built based on the tester's own experience or that of other people, available defect and failure data, and from common knowledge about why software fails. This way of trying to force specific types of fault to occur is sometimes called an 'attack' or 'fault attack'. Refer to Whittaker [2002, 2003] and Andrews and Whittaker [2006].

4.4.2 Exploratory testing

Exploratory testing An approach to testing in which the testers dynamically design and execute tests based on their knowledge, exploration of the test item and the results of previous tests.

Exploratory testing is a hands-on approach in which testers are involved in minimum planning and maximum test execution. The planning usually involves the creation of a test charter, a short declaration of the scope of a short (one- to two-hour) time-boxed test effort, the objectives and possible approaches to be used.

The test design and test execution activities are performed in parallel, typically without formally documenting the test conditions, test cases or test scripts. The tests are informal, because they are not pre-defined or documented in advance in detail (although a test charter is most often written in advance). This does not mean that other, more formal testing techniques will not be used. For example, the tester may decide to use BVA, but will think through and test the most important boundary values without necessarily writing them down. Some notes will be written while the exploratory testing is going on, so that a report can be produced afterwards.

One typical way to organize and manage exploratory testing is to have sessions, hence this is also known as session-based testing. Each session is time-boxed, for example with a firm time limit of 90 minutes. A test charter will give a list of test conditions (sometimes referred to as objectives for the test session), but the testing does not have to conform completely to that charter, particularly if new areas of high risk are discovered in the session. The test session is completely dedicated to the testing, without extraneous interruptions.

Test logging is undertaken as test execution is performed, documenting (at a high level) the key aspects of what is tested, any defects found and any thoughts about possible further testing, possibly using test session sheets. A key aspect of exploratory testing is learning: learning by the tester about the software, its use, its strengths and its weaknesses. As its name implies, exploratory testing is about exploring, finding out about the software, what it does, what it does not do, what works and what does not work. The tester is constantly making decisions about what to test next and where to spend the (limited) time.

At the end of a session, there is usually a debriefing session with stakeholders. The test charter or high-level test objectives would be considered as high-level test conditions or coverage items, so that stakeholders can see what has been tested in the session.

This is an approach that is most useful when there are no or poor specifications and when time is severely limited. It can also serve to complement other, more formal testing, helping to establish greater confidence in the software. In this way, exploratory testing can be used as a check on the formal test process by helping to ensure that the most serious defects have been found.

Exploratory testing is described in Kaner *et al.* [1999, 2002], Copeland [2003], Hendrickson [2013] and Sabourin [2024]. Other ways of testing in an exploratory way (attacks) are described by Whittaker [2003, 2009].

4.4.3 Checklist-based testing

Checklist-based testing is testing based on experience, but that experience has been summarized and documented in a checklist. Testers use the checklist to design, implement and execute tests based on the items or test conditions found in the checklist. The checklist may be based on:

> **Checklist-based testing** An experience-based test technique in which test cases are designed to exercise the items of a checklist.

● experience of the tester

● knowledge, for example what is important for the user

● understanding of why and how software fails.

When using a checklist, the tester may modify it by adding new questions or things to check or may just use what is already there. An experienced tester may be the one who writes the first version of the checklist to help less experienced testers do better testing.

Checklists can be general or, more likely, aimed at particular areas such as different test types and levels. For example, a checklist for functional testing would be different from one aimed at testing non-functional quality attributes, and a checklist for component testing would be different from one for usability testing.

If different people use the same checklist, it is more likely that there will be a degree of consistency in what is tested. However, the actual tests executed may be quite different since the checklist is only mentioning high-level items. This variability, even while using the same checklist, may help to uncover more defects and achieve different levels of coverage (if that is measured), even though it is less repeatable due to human creativity (which is a good thing).

Checklists should not be aimed at things that are better checked by tools, such as static analysis tools, or items that are too general. Many of the best checklist questions are devised after a defect was missed by testing. Putting a question on the checklist helps to ensure that same mistake won't happen again. The convention is that each item on the checklist is a question, and it is good practice to have consistency in the way the question is answered. For example, 'no' always means there is a problem and 'yes' means it's ok.

It is also important to maintain the checklists over time. Some checklist questions may no longer be relevant (a previously common defect no longer occurs due to better use of static analysis, for example). It is tempting to keep adding more questions, but the best checklists are kept to a manageable size (e.g. one page). This means that the less useful questions need to be weeded out, and this should be done regularly.

A checklist is not a substitute for entry and exit criteria (although they may also have their own checklists). Checklists may include questions for requirements or user stories, graphical interface properties, quality characteristics, test conditions or other relevant things.

Refer to Brykczynski [1999], Nielsen [1994] for usability checklists and Gawande [2009] for more on checklist-based testing.

4.5 COLLABORATION-BASED TEST APPROACHES

In this section we will look at three approaches to testing which have become more prominent in Agile development. All of these approaches may use any of the test techniques we have already discussed in this chapter. The techniques are focused on detecting different types of defects. These approaches are focused on avoiding defects by better communication and collaboration between developers, testers and stakeholders.

The emphasis is on collaboration, working together with other people. Good communication is the way in which people can understand each other and work together to create the best software or system for the users.

In this section, look for the definitions of the Glossary terms **acceptance criteria** and **acceptance test-driven development** (also referred to by its acronym ATDD).

4.5.1 Collaborative user story writing

What is a user story? It is a short and simple description of a feature which someone wants in the system being developed, written from their perspective. For example, 'a user can find the price of an item without having to sign in', 'a user can order up to five tickets at a time', or 'a user can search for classical choral music concerts by location and date'.

User stories are often expressed in a slightly more formal way using the format 'As a <role>, I want <goal to be achieved> so that I can <resulting business value>', and this may then include acceptance criteria (discussed in Section 4.5.2). The user story can also include non-functional criteria. For example, 'As a concert-goer, I want the list of classical choral music concerts near me to be displayed within 2 seconds so that I can quickly buy a ticket for a particular date'.

User stories are best written by the users, customers or product owners of the system, or by a business representative or stakeholder, but developers and testers may help to clarify the story, particularly for acceptance criteria. User stories may be written as a collaborative effort, using brainstorming techniques and mind-mapping diagrams. Working together, the user stories should produce a better shared vision of what is wanted, taking into account business, development and testing concerns.

A user story may not spring into being fully formed but may develop over time. Collaboration is important to help everyone to have a common understanding. Frequent informal reviews also help to refine user stories over time. Refinement sessions are held to review, update and clean up the product backlog, including adding detail to user stories and splitting them if they are becoming too large.

Because user stories often start by being written on small cards (or Post-it notes), Ron Jeffries [Jeffries *et al.* 2000] describes three aspects of user story development as the 'Three C's':

- *Card*: The card is the physical media on which a user story is written. It identifies the requirement, its criticality, expected development and test duration, and the acceptance criteria for that story. The description has to be accurate, as it will be used in the product backlog.

- *Conversation*: The conversation explains how the software will be used. The conversation can be documented or verbal. Testers, having a different point of view than developers and business representatives, bring valuable input to the exchange of thoughts, opinions and experiences. Conversation begins during the release-planning phase and continues as the story is implemented.

- *Confirmation*: The acceptance criteria, discussed in the conversation, are used to confirm that the story is done. Both positive and negative tests should be used to cover the criteria. During confirmation, various participants play the role of a tester. These can include developers as well as specialists focused on performance, security, interoperability and other quality characteristics. To confirm a story has been completed, the defined acceptance criteria should be tested and shown to be satisfied. Note that this is something that testers are typically very good at, challenging assumptions and thinking about 'what if' scenarios and what could go wrong.

  One way to recognize the characteristics of a good user story is by using the criteria from the acronym INVEST:

- *Independent*: Stories that are independent are much easier to work with. Dependent stories can be harder to estimate and may lead to prioritization and planning issues. For example, if a high priority story is dependent on a lower priority story, which is implemented first?

- *Negotiable*: A good story is negotiable. That is, there are not a lot of details given thereby necessitating further discussion as the story is implemented. Having a lot of detail written down (as notes of conversations) can give the impression that everything is known about the story, and this may supress further discussion that would otherwise highlight additional important information.

  When details are known, they should be turned into tests. This leaves the notes to serve as open questions that can be answered during further conversations.

- *Valuable*: Every story needs to be valuable to the customer, meaning an end-user or a purchaser of the software. Stories that are only valuable to developers probably should be reworded so that they express the benefit to the customer. For example, 'The system shall implement load balancing' could be reworded along the lines of 'The system will make optimal use of the system resources for maximum efficiency'.

- *Estimate-able*: A good story can be estimated. Mike Cohn [2009] highlights three reasons for stories being difficult/impossible to estimate: developers lack domain knowledge, developers lack technical knowledge, the story is too big. Where developers lack domain knowledge, they need to discuss the story with the customers to gain a better understanding of it. Where they lack technical knowledge, they need to spend a little time (by way of a separate story) doing research or experimenting to gain sufficient knowledge to allow a reasonable estimate to be made. Where the story is too big, it should be broken down into a number of smaller stories.

- *Small*: Good stories tend to be small, but not too small. Stories can be too small if, for example, writing it down and estimating will take longer than the actual implementation, such as with a minor bug fix. In this case a number of very small stories can be combined into one story of reasonable size.

  An epic is a high-level description of development work which is too large to complete in a single sprint. An epic may contain multiple stories, some of which may be very complex. Compound stories can typically be broken down into a number of smaller stories as the details of the compound story are discussed and refined. Complex stories, on the other hand, are more difficult to break down because of uncertainty—a lack of domain or technical knowledge. In these situations, complex stories can be divided into two stories, the first of which undertakes research or experimentation to gain sufficient knowledge to allow the complex user story to proceed.

- *Testable*: A good story is testable. If a story cannot be tested, then the developers can never know when they have finished implementing a story. If a stakeholder doesn't know how to test a story, how can they tell if it has been implemented correctly? If a user story is not clear or specific enough, or if the business value is not there or not clear enough, then that story is not testable. Sometimes a stakeholder may need the help of the testers to make the user stories testable; this collaboration should be encouraged.

  Stories that are untestable may be non-functional stories. For example, 'The user will be able to complete the booking quickly'. This does not define what is meant by 'quickly' so the story must be rewritten, or a testable acceptance criteria added, to make it testable.

4.5.2 Acceptance criteria

Acceptance criteria The criteria that a component or system must satisfy in order to be accepted by a user, customer or other authorized entity.

Acceptance criteria are the way in which the software system is evaluated by those who will be using it. They are test conditions for a user story; if they pass, then that part of the system which is being tested is deemed to be what the user or customer expected. It is important that these expectations are documented (as acceptance criteria). If they are not, a user's expectation may be very different from that of a developer implementing a particular feature. Acceptance criteria typically dig into the user story and make it more tangible.

Setting acceptance criteria should also be done in collaboration with users and other stakeholders, to make sure that they do represent their expectations, and so that developers and testers know what is expected to be built and how it will be tested. Hence, they are formulated during the Conversation part of the Three C's.

The purposes of acceptance criteria are:

- To define the scope of the user story.
- To reach consensus among all the stakeholders about what should be developed.
- To describe both positive and negative scenarios, i.e. what the system should do and what it should not do.
- To serve as a basis for the user story acceptance testing, such as ATDD (refer to Section 4.5.3).
- To support accurate estimation and planning of the work.

The way in which acceptance criteria are documented can vary, depending on the organization or team. As long as they are well-defined and unambiguous, the format can be whatever works.

There are two common formats for acceptance criteria:

- Scenario-oriented, such as 'Given—When—Then' format used in Behaviour-Driven Development (BDD), also known as the Gherkin language and described in Section 2.1.3.
- Rule-oriented, where the acceptance criteria are listed, e.g. in a bullet list, or as a tabulated form of input-output mapping.

4.5.3 Acceptance test-driven development (ATDD)

Acceptance test-driven development is a way of developing software by writing the acceptance tests before implementing the user stories or writing the code. It is one of several test-first approaches described in Section 2.1.3. ATDD is not a test technique as described in Sections 4.2 to 4.4, but ATDD can use any of those techniques to write the acceptance tests.

The acceptance tests are created collaboratively by team members, such as developers, testers and customers or users. They come together (in person or using an internet collaboration tool such as Zoom or Teams), in a test specification workshop. A starting point would be any existing user stories and/or acceptance criteria, but even if they are used, they will often be refined and changed as they are analyzed from different perspectives and discussed. The agreed user stories and acceptance criteria are documented as a product of the workshop. The workshop gives everyone a chance to resolve ambiguities or misunderstandings, and to identify defects in the user stories or acceptance criteria.

The next step is to create the specific test cases. This may be done collaboratively in a continued workshop, or it could be done later. It could also be a task for the tester(s) to write the test cases from the acceptance criteria. The test cases, with specific data and specified transactions, provide examples of how the system is expected to work. A test is a particular example of how the system should work; a similar approach is described by Adzic [2009] as 'specification by example'.

Making things specific often reveals further misunderstandings or ambiguities, so the test cases should be reviewed by the team (including developers and business stakeholders). Having the test cases pre-defined helps the developers to produce

Acceptance test-driven development A collaboration-based test-first approach that defines acceptance tests in the stakeholders' domain language.

code that will pass those tests and thereby implement what was wanted. The test cases created can be manual tests or automated tests.

Normally, positive tests are written first—the 'happy path' where we are confirming that things are working as expected, without worrying about exceptions or error conditions. But negative tests (testing that the system doesn't do what it shouldn't) is also important to include. This includes exceptions from the main path and testing that the correct error messages are shown as required.

However, both positive and negative tests are testing the functionality of the system. Non-functional quality characteristics should also have acceptance criteria and tests to determine how well the system works (including performance efficiency, usability, security testing and others).

Because the test cases need to be understood by all stakeholders, they should be written in a way that everyone can understand them. They could be written in natural language for ease of understanding, but they still need to include the elements of the test, such as pre-conditions, inputs, expected results and post-conditions. However, the format being used also depends on whether the tests will be automated or not. When they are automated a more formal language will be used for documenting them.

As with all good test cases, each test case should have a single clear objective. Tests that cover the same characteristics as another test may be an unneeded duplication. The tests should cover all of the characteristics described in a user story and in the acceptance criteria and should not go beyond what has been specified.

If the acceptance test cases are recorded in a form that can be used by a test execution tool, then the developers can ensure that the tests are implemented in that tool, writing any support code that may be needed, as they are implementing the code for the user story. If the acceptance tests are automated, they are sometimes referred to as executable requirements.

ATDD example

We will show an example of using ATDD to design some functional test cases for a Book Club. The Book Club is online, and members are allowed to download the book of the month and are also allowed (and encouraged) to leave reviews of the books they have read. We will revisit this example later on for the chapter exercise, looking at different acceptance criteria and test cases.

As a member of the Book Club, I want to be able to read other members' reviews in full from any book we have read in the past three years. A Book Club Admin can delete reviews that are inappropriate or offensive, thereby terminating their membership.

　Acceptance criteria

AC1: A member can see any other member's review for any book in the past three years.

AC2: Book Club Admin can delete any review.

AC3: A member whose review has been deleted is no longer a member and cannot download books or leave any reviews.

We will look at a first step in developing test cases for each of these acceptance criteria. Actually, we will be using high-level test conditions rather than specific test data and test inputs.

In the tests below, 'book' means the current book of the month for the current date. Unless otherwise stated, the current date is 30 March 2024. The starting point for each test is a screen showing the current book of the month, with author, title and ISBN number (and photo of the book cover if available).

> AC1: A member can see any other member's review for any book in the past three years.

Test case reference	Input	Expected outcome
1	Member requests to see reviews for the current book.	March 2024's book reviews are displayed.
2	Member requests to see reviews for the book from April 2021.	April 2021's book reviews are displayed.
3	Member requests to see reviews for the book from March 2021.	Error message: reviews are only available for 3 years.

> AC2: Book Club Admin can delete any review.

The following tests must be done in order. The pre-condition is that there is at least a review for February 2024's book from member X.

Test case reference	Input	Expected outcome
4	Member (not Admin) requests to see reviews for January 2024.	List of n reviews from January 2024 is displayed, including one from member X. (n is 1 or more).
5	Admin finds member X's review of the book for January 2024 and deletes the review.	Review is deleted.
6	After Test 5, member (not Admin) requests to see reviews for January 2024.	List of n-1 reviews from January 2024 is displayed, with no review from member X.

> AC3: A member whose review has been deleted is no longer a member, and cannot download books or leave any reviews.

Test case reference	Input	Expected outcome
7	Member X requests to download this month's book.	Error message: only members are allowed to download books.
8	Member X submits a review for a book that was previously downloaded (while a member).	Error message: only members are allowed to review books.

We may have some questions after thinking of these test cases, such as:

- What about reviews from more than three years ago—are they archived? Can Admin see them?
- Can a review be deleted from more than three years ago? If so, is the membership terminated?
- Can a member whose review has been deleted join again, or are they barred for life?
- Does Admin have sole control over deletions, or are deletions done in consultation with other members? Who decides what is offensive or inappropriate? Does the reason for deletion need to be communicated to the offending member? Is a note left on the website to say that a review has been deleted?

These questions may then be used to further refine the user stories and would lead to other acceptance criteria.

4.6 CHOOSING TEST TECHNIQUES (ADDITIONAL MATERIAL)

In this optional extra section, we will look at the factors that go into the decision about which techniques to use when.

Which technique is best? This is the wrong question! Each technique is good for certain things, and not as good for other things. For example, one of the benefits of white-box techniques is that they can find things in the code that are not supposed to be there, such as Trojan Horses or other malicious code. However, if there are parts of the specification that are missing from the code, black-box techniques can find that. White-box techniques can only test what is there. If there are things missing both from the specification and from the code, then only experience-based techniques would find them. Collaboration-based test approaches help to make all techniques more effective. Each individual technique is aimed at particular types of defect as well. For example, state transition testing is unlikely to find boundary defects.

The choice of which test technique to use depends on a number of factors, which we discuss below.

Some techniques are more applicable to certain situations and test levels; others are applicable to all test levels. The best testing uses a combination of test techniques.

This chapter has covered the most popular and commonly used software test techniques. There are many others that fall outside the scope of the Syllabus that this book is based on. With so many testing techniques to choose from, how are testers to decide which ones to use?

The use of techniques can vary in formality, from very informal (with little or no documentation) to very formal, where information about test conditions and why particular techniques are used is recorded. The level of formality also depends on other factors as discussed below. For example, safety or regulatory industries require a higher level of formality. The maturity of the organization, the life cycle model being used, and the knowledge and skills of the testers also influence the level of formality.

Perhaps the single most important thing to understand is that the best test technique is no single test technique. Because each test technique is good at finding one specific class of defect, using just one technique will help ensure that many (perhaps most but not all) defects of that particular class are found. Unfortunately, it may also help to ensure that many defects of other classes are missed! Using a variety of techniques will therefore help ensure that a variety of defects are found, resulting in more effective testing.

So how can we choose the most appropriate test techniques to use? The decision will be based on a number of factors, both internal and external.

The factors that influence the decision about which technique to use are:

- *Type of component or system*: The type of component (for example, embedded, graphical, financial, etc.) will influence the choice of techniques. For example, a financial application involving many calculations would benefit from boundary value analysis.

- *Component or system complexity*: More complex components or systems are likely to have more defects, and defects may be harder to find. Using a different technique to address aspects of that complexity will give better defect detection from the testing. For example, a simple field with numerical input would be a good candidate for equivalence partitioning and boundary value analysis, but a screen with many fields, with complex dependencies, calculations and validation rules depending on aspects that change over time, would benefit from additional techniques such as decision table testing, state transition testing and white-box test techniques.

- *Regulatory standards*: Some industries have regulatory standards or guidelines that govern the test techniques used. For example, the aircraft industry requires the use of equivalence partitioning, boundary value analysis and state transition testing for high integrity systems, together with statement, branch or decision coverage or modified condition decision coverage depending on the level of software integrity required.

- *Customer or contractual requirements*: Sometimes contracts specify particular test techniques to use (most commonly statement or branch coverage).

- *Risk levels and risk types*: The greater the risk level (for example, safety-critical systems), the greater the need for more thorough, formal testing. Commercial risk may be influenced by quality issues (so more thorough testing would be appropriate) or by time to market issues (so exploratory testing would be a more appropriate choice). Risk type might, for example, tell us that a risk is related to usability, performance, security or functionality. Of course, the correct test technique needs to be chosen that addresses the risk type that is being mitigated.

- *Test objectives*: If the test objective is simply to gain confidence that the software will handle typical data in a contact form, then equivalence partitioning would be a sensible approach. If the objective is for very thorough testing, then more rigorous and detailed techniques (including white-box test techniques) should be chosen.

- *Available documentation*: Whether or not documentation (for example, a work product such as a requirements specification) exists, and how up-to-date it is, will affect the choice of test techniques. The content and style of the work products will also influence the choice of techniques (for example, if decision tables or state graphs have been used, then the associated test techniques should be used).

- *Tester knowledge and skills*: How much testers know about the system and about test techniques will clearly influence their choice of techniques. Experience-based techniques are particularly based on tester knowledge and skills.

- *Available tools*: If tools are available for a particular technique, then that technique may be a good choice. Since test techniques are based on models, the models available (that is, developed and used during the specification, design and implementation of the system) will to some extent govern which test techniques can be used. For example, if the specification contains a state transition diagram, state transition testing would be a good technique to use.

- *Time and budget*: Ultimately, how much time there is available will always affect the choice of test techniques. When more time is available, we can afford to select more techniques. When time is severely limited, we will be limited to those that we know have a good chance of helping us find just the most important defects.

- *Software development life cycle model*: A sequential life cycle model will lend itself to the use of more formal techniques, whereas an iterative life cycle model may be better suited to using an exploratory test approach.

- *Expected use of the software*: If the software is to be used in safety-critical situations, for example medical monitoring devices or car-driving technology, then the testing should be more thorough, and more techniques should be chosen.

- *Previous experience with using the test techniques on the component or system to be tested*: Testers tend to use techniques that they are familiar with and more skilled at, rather than less familiar techniques. With this familiarity, you can become very effective at finding similar defects to those that have occurred before. But be aware that you may get better results from using a technique that is less familiar, and when you do use it, you will increase your skill and familiarity with it.

- *The types of defects expected in the component or system*: Knowledge of the likely defects will be very helpful in choosing test techniques (since each technique is good at finding a particular type of defect). This knowledge could be gained through experience of testing a previous version of the system and previous levels of testing on the current version.

It is important to remember that intelligent, experienced testers see test techniques (and indeed test strategies, which we'll discuss in Chapter 5) as tools to be employed wherever needed and useful. You should use whatever techniques and strategies, in whatever combinations, make sense to ensure adequate coverage of the system under test, and achievement of the objectives of testing. Feel free to combine the test techniques discussed in this chapter with whatever inspiration you have, along with process-related, rule-based and data-driven techniques. Use your brain and do what makes sense.

CHAPTER REVIEW

Let's review what you have learned in this chapter.

From Section 4.1 (test techniques overview), you should be able to give reasons why black-box, white-box and experience-based approaches are useful, and be able to explain the characteristics and differences between these types of techniques. You should know that black-box techniques are related to behaviour, white-box are related to structure and experience-based are related to testers' knowledge. You should know the Glossary terms **black-box test technique**, **coverage**, **coverage item**, **experience-based test technique**, **test technique** and **white-box test technique**.

From Section 4.2, you should be able to write test cases from given software models using equivalence partitioning (EP), Boundary Value Analysis (BVA), decision table testing and state transition testing. You should understand and be able to apply each of these four techniques, understand what level and type of testing could use each technique and how coverage can be measured for each of them. You should know the Glossary terms **boundary value analysis**, **decision table testing**, **equivalence partitioning** and **state transition testing**.

From Section 4.3, you should be able to describe the concept and importance of code coverage. You should be able to explain the concepts of statement and branch coverage and understand that these concepts can also be used at test levels other than component testing (such as business procedures at system test level). You should be able to write test cases from given control flows using statement testing and branch testing, and you should be able to assess statement and branch coverage for completeness. You should know the Glossary terms **branch coverage**, **coverage** and **statement coverage**.

From Section 4.4, you should be able to explain the reasons for writing test cases based on intuition, experience and knowledge about common defects and you should be able to compare experience-based techniques with black-box test techniques. You should know the Glossary terms **checklist-based testing**, **error guessing** and **exploratory testing**.

From Section 4.5, you should be able to describe collaboration-based test approaches, which apply test techniques when working with other people. You should be able to explain what user stories are and how to write them using the 3 C's (Card, Conversation and Confirmation). You should be able to describe what acceptance criteria are used for and how they can be written. You should be able to write test cases using ATDD. You should know the Glossary terms **acceptance criteria** and **acceptance test-driven development**.

SAMPLE EXAM QUESTIONS

Question 1 Which of the following statements about checklist-based testing is true?

a. A checklist contains test conditions and detailed test cases and procedures.

b. A checklist can be used for functional testing but not for non-functional testing.

c. Checklists should always be used exactly as written and should not be modified by the tester.

d. Checklists may be based on experience of why and how software fails.

Question 2 In a competition, ribbons are awarded as follows: less than 12 metres, no ribbon; a yellow ribbon up to 25 metres; a red ribbon up to 35 metres; and a blue ribbon for further than that.
   What distances (in metres) would be chosen using BVA?

a. 0, 11, 12, 25, 26, 35, 36.

b. 11, 12, 13, 29, 30, 31, 40.

c. 7, 18, 32, 39.

d. 0, 12, 13, 26, 27, 36, 37.

Question 3 Which statement about tests based on increasing statement or branch coverage is true?

a. Increasing statement coverage may find defects where other tests have not taken both True and False conditional branches from a decision statement.

b. Increasing statement coverage may find defects in code that was exercised by other tests.

c. Increasing branch coverage may find defects where other tests have not taken both True and False conditional branches from a decision statement.

d. Increasing branch coverage may find defects in code that was exercised by other tests.

Question 4 Why are both black-box and white-box test techniques useful?

a. They find different types of defect.

b. Using more techniques is always better.

c. Both find the same types of defect.

d. Because specifications tend to be unstructured.

Question 5 What does branch coverage measure?

a. That the branches from a decision statement have been exercised.

b. That both conditional and unconditional branches have been exercised.

c. That statements and unconditional branches have been exercised.

d. That changes by developers to the code branching structure have been exercised.

Question 6 Which of the following would be an example of decision table testing for a financial application, applied at the system-test level?

a. A table containing rules for combinations of inputs to two fields on a screen.

b. A table containing rules for interfaces between components.

c. A table containing rules for mortgage applications.

d. A table containing rules for basic arithmetic to two decimal places.

Question 7 Which of the following could be a coverage measure for state transition testing?

V All states have been reached.

W The response time for each transaction is adequate.

X Every valid transition has been exercised.

Y All boundaries have been exercised.

Z All transitions have been exercised.

a. X, Y and Z.

b. V, X, Y and Z.

c. W, X and Y.

d. V, X and Z.

Question 8 Postal rates for light letters are $0.25 up to 10g, $0.35 up to 50g, plus an extra $0.10 for each additional 25g up to 100g.
   Which test inputs (in grams) would be selected using equivalence partitioning (EP)?

a. 8, 42, 82, 102.

b. 4, 15, 65, 92, 159.

c. 10, 50, 75, 100.

d. 5, 20, 40, 60, 80.

Question 9 Which of the following could be used to assess the coverage achieved for the use of black-box test techniques?

V Branches exercised.

W Partitions exercised.

X Boundaries exercised.

Y State transitions exercised.

Z Statements exercised.

a. V, W, Y or Z.

b. W, X or Y.

c. V, X or Z.

d. W, X, Y or Z.

Question 10 Which of the following would white-box test techniques be most likely to be applied to?

1. Boundaries between mortgage interest rate bands.

2. An invalid transition between two different arrears statuses.

3. The business process flow for mortgage approval.

4. Control flow of the program to calculate repayments.

a. 2, 3 and 4.

b. 2 and 4.

c. 3 and 4.

d. 1, 2 and 3.

Question 11 Acceptance criteria are used for which of the following?

P Describing both positive and negative scenarios.

Q Serving as the basis for a user story.

R Reaching a consensus among stakeholders.

S Ensuring that the scope of the system being developed is fully described.

T Confirming that estimates have been accurate by checking time spent on the story.

a. P, Q and R.

b. Q, S and T.

c. P, Q and S.

d. R, S and T.

Question 12 Which of the following statements about the relationship between statement coverage and branch coverage is correct?

a. 100% branch coverage is achieved if statement coverage is greater than 90%.

b. 100% statement coverage is achieved if branch coverage is greater than 90%.

c. 100% branch coverage always means 100% statement coverage.

d. 100% statement coverage always means 100% branch coverage.

Question 13 Why are experience-based test techniques good to use?

a. They can find defects missed by black-box and white-box test techniques.

b. They do not require any training to be as effective as formal techniques.

c. They can be used most effectively when there are good specifications.

d. They will ensure that all of the code or system is tested.

Question 14 If you are flying with an economy ticket, there is a possibility that you may get upgraded to business class, especially if you hold a gold card in the airline's frequent flier program. If you do not hold a gold card, there is a possibility that you will get bumped off the flight if it is full and you check in late. This is shown in Figure 4.4. Note that each box (that is, statement) has been numbered.

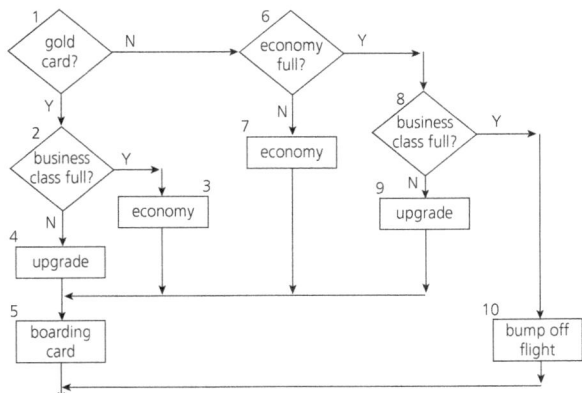

FIGURE 4.4 Control flow diagram for flight check-in

Three tests have been run:

Test 1: Gold card holder who gets upgraded to business class.

Test 2: Non-gold card holder who stays in economy.

Test 3: A person who is bumped from the flight.
What is the statement coverage of these three tests?

a. 60%.

b. 70%.

c. 80%.

d. 90%.

Question 15 Consider the control flow shown in Figure 4.5. In this example, if someone is a member, then they get a 10% discount, but only on items with an ItemCode of 25 or less. The following tests have already been run:

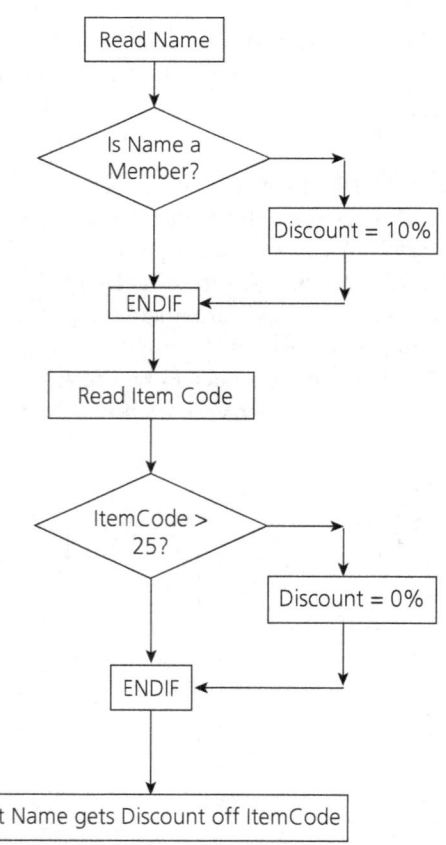

FIGURE 4.5 Control flow diagram for Question 15

Test 1: Name is a member, ItemCode = 50

Test 2: Name is not a member, ItemCode = 27
What is the statement coverage and branch coverage of these tests?

a. Statement coverage = 100%, Branch coverage = 100%.

b. Statement coverage = 100%, Branch coverage = 80%.

c. Statement coverage = 90%, Branch coverage = 100%.

d. Statement coverage = 100%, Branch coverage = 90%.

Question 16 Which of the following are associated with exploratory testing?

   V The use of other test techniques, such as EP/ BVA.

   W Fault attacks.

   X A test charter.

   Y Simultaneous test design and execution.

   Z Needs consensus of all stakeholders.

a. All of them.

b. V, X and Y.

c. V, W, X and Z.

d. W, Y and Z.

Question 17 Given the state diagram in Figure 4.6, which test case is the minimum series of valid transitions to cover every state?

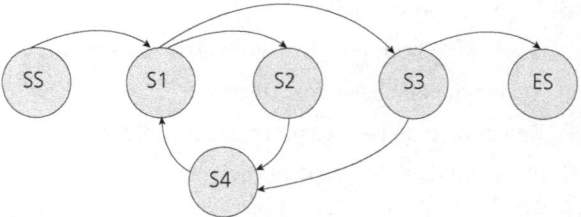

FIGURE 4.6 State diagram for PIN entry

a. SS – S1 – S2 – S4 – S1 – S3 – ES.

b. SS – S1 – S2 – S3 – S4 – ES.

c. SS – S1 – S2 – S4 – S1 – S3 – S4 – S1 – S3 – ES.

d. SS – S1 – S4 – S2 – S1 – S3 – ES.

Question 18 Which of the following is NOT a critical aspect of a user story?

a. Interaction between users, developers and testers.

b. Estimation of the time to develop the feature.

c. Documentation describing the business value.

d. Conditions that must be met to satisfy stakeholders.

Question 19 You are using ATDD to design test cases for the following user story:

As a member of the Book Club, I want to be able to access each month's book and leave a review, so that I can share my views with other Book Club members.
Acceptance criteria:
AC1: A member can download the book of the month.
AC2: A non-member can see the author and title of the book but cannot download it.
AC3: A member can leave a review for a book from two days after downloading it until the end of the following month.

Which of the following test cases are the best ones to test AC3?

J A member downloads the book on 30 January and leaves a review for it on 1 February.

K A non-member reads the book of the month but is prevented from leaving a review.

L The author and title of the book of the month change on the first of each month.

M A member downloads the book on 28 February and is prevented from leaving a review on 1 April.

N A member downloads the book on 30 January and is prevented from leaving a review on 31 January.

a. J, K and L.

b. J, K and N.

c. J, M and N.

d. K, M and N.

EXERCISES

Exercises based on the techniques covered in this chapter are given in this section. Worked solutions are given in the next section.

Equivalence Partitioning/Boundary Value Analysis exercise

Scenario: If you take the train before 9:30 am or in the afternoon after 4:00 pm until 7:30 pm (the rush hour), you must pay full fare. A saver ticket is available for trains between 9:30 am and 4:00 pm, and after 7:30 pm.

What are the partitions and boundary values to test the train times for ticket types? (Use two-value BVA.) Which are valid partitions, and which are invalid partitions? What are the boundary values? (A table may be helpful to organize your partitions and boundaries.) Derive test cases for the partitions and boundaries.

Are there any questions you have about this requirement? Is anything unclear?

Decision table exercise

Scenario: If you hold an over 60s rail card, you get a 34% discount on whatever ticket you buy. If you are travelling with a child (under 16), you can get a 50% discount on any ticket if you hold a family rail card, otherwise you get a 10% discount. You can only hold one type of rail card.

Produce a decision table showing all the combinations of fare types and resulting discounts and derive test cases from the decision table.

State transition exercise

Scenario: A website shopping basket starts out as empty. As purchases are selected, they are added to the shopping basket. Items can also be removed from the shopping basket. When the customer decides to check out, a summary of the items in the basket and the total cost are shown, for the customer to say whether this is OK or not. If the contents and price are OK, then you leave the summary display and go to the payment system. Otherwise, you go back to shopping (so you can remove items if you want).

a. Produce a state diagram showing the different states and transitions. Define a test, in terms of the sequence of states, to cover all transitions.

b. Produce a state table. Give an example test for an invalid transition.

Statement and branch testing exercise

Scenario: A vending machine dispenses either hot or cold drinks. If you choose a hot drink (for example, tea or coffee), it asks if you want milk (and adds milk if required), then it asks if you want sugar (and adds sugar if required), then your drink is dispensed.

a. Draw a control flow diagram for this example. (Hint: regard the selection of the type of drink as one statement.)

b. Given the following tests, what is the statement coverage achieved? What is the branch coverage achieved?

Test 1: Cold drink.
Test 2: Hot drink with milk and sugar.

c. What additional tests would be needed to achieve 100% statement coverage? What additional tests would be needed to achieve 100% branch coverage?

ATDD exercise

You are using ATDD to design test cases for the following user story:

As a member of the Book Club, I want to be able to access each month's
book and leave a review, so that I can share my views with other
Book Club members.

Acceptance criteria:

AC1: A member can download the book of the month.

AC2: A member can leave a review for a book from the day after
downloading it.

AC3: Members can read any review of any book.

AC4: Non-members can read the first paragraph of the reviews for the
current month.

Write one or two test cases for each of the acceptance criteria (the test cases do not
need to have detailed input but should cover relevant test conditions).

EXERCISE SOLUTIONS

EP/BVA exercise

The first thing to do is to establish exactly what the boundaries are between the full fare and saver fare. Let's put these in a table to organize our thoughts:

Scheduled departure time	≤ 9:29 am	9:30 am – 4:00 pm	4:01 pm – 7:30 pm	≥ 7:31 pm
Ticket type	full	saver	full	saver

We have assumed that the boundary values are: 9:29 am, 9:30 am, 4:00 pm, 4:01 pm, 7:30 pm and 7:31 pm. By setting out exactly what we think is meant by the specification, we may highlight some ambiguities or, at least, raise some questions. This is one of the benefits of using the technique! For example:

'When does the morning rush hour start? At midnight? At 11:30 pm the previous day? At the time of the first train of the day? If so, when is the first train? 5:00 am?'

This is a rather important omission from the specification. We could make an assumption about when it starts, but it would be better to find out what is correct.

- If a train is due to leave at exactly 4:00 pm, is a saver ticket still valid?
- What if a train is due to leave before 4:00 pm but is delayed until after 4:00 pm? Is a saver ticket still valid? (That is, if the actual departure time is different from the scheduled departure time.)

Our table has helped us to see where the partitions are. All of the partitions in the table above are valid partitions. It may be that an invalid partition would be a time that no train was running, for example before 5:00 am, but our specification did not mention that! However, it would be good to show this possibility also. We could be a bit more formal by listing all valid and invalid partitions and boundaries in a table, as we described in Section 4.2.1, but in this case, it does not actually add a lot, since all partitions are valid.

Here are the test cases we can derive for this example:

Test case reference	Input	Expected outcome
1	Depart 4:30 am	Pay full fare
2	Depart 9:29 am	Pay full fare
3	Depart 9:30 am	Buy saver ticket
4	Depart 11:37 am	Buy saver ticket
5	Depart 4:00 pm	Buy saver ticket
6	Depart 4:01 pm	Pay full fare
7	Depart 5:55 pm	Pay full fare
8	Depart 7:30 pm	Pay full fare
9	Depart 7:31 pm	Buy saver ticket
10	Depart 10:05 pm	Buy saver ticket

Note that test cases 1, 4, 7 and 10 are based on equivalence partition values; test cases 2, 3, 5, 6, 8 and 9 are based on boundary values. There may also be other information about the test cases, such as preconditions, that we have not shown here.

Decision table exercise

The fare types mentioned are an over 60s rail card, a family rail card and whether you are travelling with a child or not. With three conditions or causes, we have eight columns in our decision table below.

Causes (inputs)	R1	R2	R3	R4	R5	R6	R7	R8
over 60s rail card?	Y	Y	Y	Y	N	N	N	N
family rail card?	Y	Y	N	N	Y	Y	N	N
child also travelling?	Y	N	Y	N	Y	N	Y	N
Effects (outputs)								
Discount (%)	X/?/50%	X/?/34%	34%	34%	50%	0%	10%	0%

When we come to fill in the effects, we may find this a bit more difficult. For the first two rules, for example, what should the output be? Is it an X because holding more than one rail card should not be possible? The specification does not actually say what happens if someone does hold more than one card, that is, it has not specified the output, so perhaps we should put a question mark in this column. Of course, if someone does hold two rail cards, they probably would not admit this, and perhaps they would claim the 50% discount with their family rail card if they are travelling with a child, so perhaps we should put 50% for Rule 1 and 34% for Rule 2 in this column. Our notation shows that we do not know what the expected outcome should be for these rules!

This highlights the fact that our natural language (English) specification is not very clear as to what the effects should actually be. A strength of this technique is that it forces greater clarity. If the answers are spelled out in a decision table, then it is clear what the effect should be. When different people come up with different answers for the outputs, then you have an unclear specification!

The word 'otherwise' in the specification is ambiguous. Does 'otherwise' mean that you always get at least a 10% discount or does it mean that if you travel with a child and an over 60s card but not a family card you get 10% and 34%? Depending on what assumption you make for the meaning of 'otherwise', you will get a different last row in your decision table.

Note that the effect or output is the same (34%) for both Rules 3 and 4. This means that our third cause (whether or not a child is also travelling) actually has no influence on the output. These columns could therefore be combined with 'do not care' as the entry for the third cause. This rationalizing of the table means we will have fewer columns and therefore fewer test cases. The reduction in test cases is based on the assumption we are making about the factor having no effect on the outcome, so a more thorough approach would be to include each column in the table.

Here is a rationalized table, where we have shown our assumptions about the first two outcomes and we have also combined Rules 6 and 8 above, since having a family rail card has no effect if you are not travelling with a child.

Causes (inputs)	R1	R2	R3	R5	R6	R7
over 60s rail card?	Y	Y	Y	N	N	N
family rail card?	Y	Y	N	Y	–	N
child also travelling?	Y	N	–	Y	N	Y
Effects (outputs)						
Discount (%)	50%	34%	34%	50%	0%	10%

Here are the test cases that we derive from this table. (If you did not rationalize (or collapse) the table, then you will have eight test cases rather than six.) Note that you would not necessarily test each column, but the table enables you to make a decision about which combinations to test and which not to test this time.

Test case reference	Input	Expected outcome
1	S. Wilkes, with over 60s rail card and family rail card, travelling with grandson Josh (age 11)	50% discount for both tickets
2	Mrs M. Davis, with over 60s rail card and family rail card, travelling alone	34% discount
3	J. Rogers, with over 60s rail card, travelling with his wife	34% discount (for J. Rogers only, not his wife)
4	S. Gray, with family rail card, travelling with their daughter Betsy	50% discount for both tickets
5	Miss Congeniality, no rail card, travelling alone	No discount
6	Joe Bloggs with no rail card, travelling with his 5-year-old niece	10% discount for both tickets

Note that we may have raised some additional issues when we designed the test cases. For example, does the discount for a rail card apply only to the traveller or to someone travelling with them? Here we have assumed that it applies to all travellers for the family rail card, but to the individual passenger only for the over 60s rail card.

State transition exercise

The state diagram is shown in Figure 4.7. The initial state (S1) is when the shopping basket is empty. When an item is added to the basket, it goes to state S2, where there are potential purchases. Any additional items added to the basket do not change the state (just the total number of things to purchase). Items can be removed, which does not change the state unless the total items ordered goes from 1 to 0. In this case, we go back to the empty basket (S1). When we want to check out, we go to the summary state (S3) for approval. If the list and prices are approved, we go to payment (S4); if not, we go back to the shopping state (possibly to remove some items to reduce the total price we have to pay). There are four states and seven transitions.

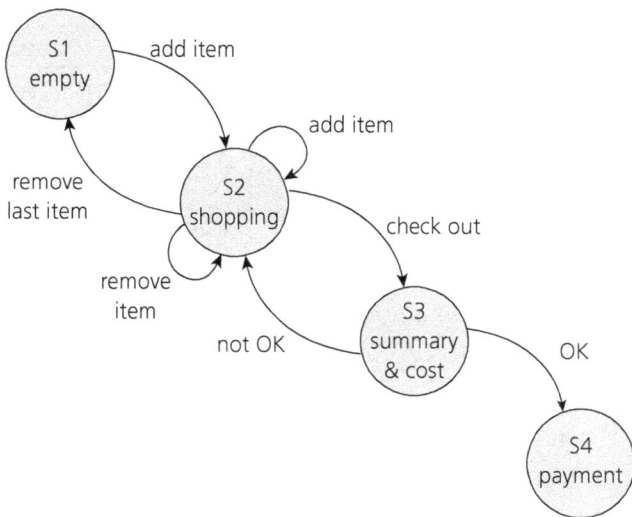

FIGURE 4.7   State diagram for shopping basket

Note that S1 is our start state for this example and S4 is the end state—this means that we are not concerned with any event that happens once we get to State S4.

Here is a test to cover all transitions. Note that the end state from one step or event is the start state for the next event, so these steps must be done in this sequence.

State	Event (action)
S1	Add item
S2	Remove (last) item
S1	Add item
S2	Add item
S2	Remove item
S2	Check out
S3	Not OK
S2	Check out
S3	OK
S4	Payment

Although our example is not interested in what happens from State 4, there would be other events and actions once we enter the payment process that could be shown by another state diagram (for example, check validity of the credit card, deduct the amount, email a receipt, etc.).

The corresponding state table is:

State or event	Add item	Remove item	Remove last item	Check out	Not OK	OK
S1 Empty	S2	–	–	–	–	–
S2 Shopping	S2	S2	S1	S3	–	–
S3 Summary	–	–	–	–	S2	S4
S4 Payment	–	–	–	–	–	–

All of the boxes that contain '–' (dash) are invalid transitions in this example. Example negative tests would include:

- Try to add an item from the summary and cost state (S3).
- Try to remove an item from the empty shopping basket (S1).
- Try to enter OK while in the shopping state (S2).

Statement and branch testing exercise

The control flow diagram is shown in Figure 4.8. Note that drawing a control diagram here illustrates that white-box testing can also be applied to the structure of general processes, not just to computer algorithms. Flowcharts are generally easier to understand than text when you are trying to describe the results of decisions taken on later events.

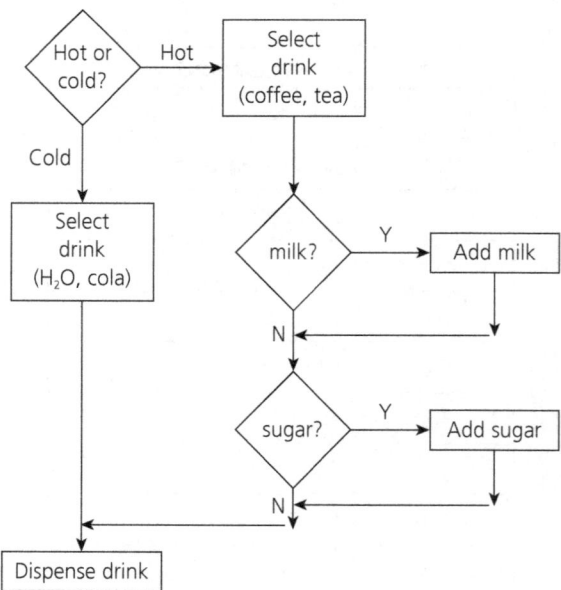

FIGURE 4.8 Control flow diagram for drinks dispenser

In Figure 4.9, we can see the route that Tests 1 and 2 have taken through our control flow graph. Test 1 has gone straight down the left-hand side to select a cold drink. Test 2 has gone to the right at each opportunity, adding both milk and sugar to a hot drink.

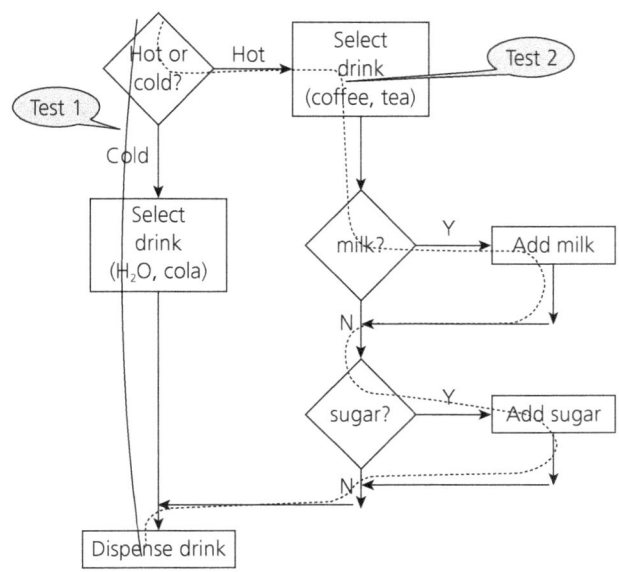

FIGURE 4.9 Control flow diagram showing coverage of tests

Every statement (represented by a box on the diagram) has been covered by our two tests, so we have 100% statement coverage.

We have not taken the No Exit from either the 'milk?' or 'sugar?' decision statements, so there are two conditional branches that we have not tested yet. We did test both of the conditional branches from the 'hot or cold?' decision statement, so we have covered four out of six decision outcomes and six out of ten branches. Branch coverage is 6/10 or 60% with the two tests.

No additional tests are needed to achieve statement coverage, as we already have 100% coverage of the statements.

One additional test is needed to achieve 100% branch coverage:

Test 3: Hot drink, no milk, no sugar.

This test will cover both of the 'No' conditional branches from the milk and sugar decision statements, so we will now have 100% branch coverage.

ATDD exercise

We show some test cases for each of the acceptance criteria in Figures 4.8 and 4.9. This is not at all an exhaustive list. Note that the inputs we suggest below are actually test conditions rather than specific values; using specific values would be better as they would then be test cases.

In the tests below, 'book' means the current book of the month for the current date. Unless otherwise stated, the current date is 30 March. The starting point for each test is a screen showing the current book of the month, with author, title and ISBN number (and photo of the book cover if available).

AC1: A member can download the book of the month.

Test case reference	Input	Expected outcome
1	Member requests download of the book.	March's book is downloaded.
2	Non-member requests download of the book.	Error message: only members are allowed to download the book.

AC2: A member can leave a review for a book from the day after downloading it.

Test case reference	Input	Expected outcome
3	Member submits a review for the book on 31 March.	Review is accepted.
4	Member submits review on 30 March.	Error message: review needs to be after the day it is downloaded.
5	Member requests last month's book. Member submits review for last month's book.	February's book is displayed, and the review is accepted.
6	Non-member submits review of the book.	Error message: only members can leave reviews.
7	Member submits review for a book they have not downloaded.	Error message: reviews can only be left for books you have downloaded.

AC3: Members can read any review of any book.

Test case reference	Input	Expected outcome
8	A member requests to see reviews of this month's book.	All reviews are displayed in full.
9	A member requests to see reviews of last month's books.	February's reviews are displayed in full.

AC4: Non-members can read the first paragraph of the reviews for the current month.

Test case reference	Input	Expected outcome
10	A non-member requests to see reviews of this month's book.	The first paragraph of each review for this month's book is displayed.
11	A non-member requests to see reviews of last months' book.	Error message: only members can see reviews of previous months' books

Note that we have not (yet) addressed any non-functional characteristics such as performance efficiency or usability. These should also be included as acceptance criteria and should also be tested.

We may also have some questions after thinking of these test cases, for example what happens if a member tries to download a book they have already downloaded? Is a one-paragraph review displayed in full for non-members? Is there a limit to how far in the past a member can request to see or submit reviews?

CHAPTER FIVE

Managing the test activities

Testing is a complex activity. It can be a distinct sub-project within the larger software development, maintenance or integration project. It usually accounts for a substantial proportion of the overall project budget. Therefore, we must understand how we should manage the testing we do.

In this chapter, we cover essential topics for test management in five sections, plus a bonus section on non-examinable but potentially valuable information. The first concerns the estimation, planning and strategizing of the test effort. The second covers the central topic of risk and how testing affects and is affected by product and project risks. The third addresses test progress monitoring, test reporting and test control. The fourth explains configuration management and its relationship to testing. The fifth section discusses the management of defects.

5.1 TEST PLANNING

> **SYLLABUS LEARNING OBJECTIVES FOR 5.1 TEST PLANNING (K3)**
>
> **FL-5.1.1** Exemplify the purpose and content of a test plan (K2)
>
> **FL-5.1.2** Recognize how a tester adds value to iteration and release planning (K1)
>
> **FL-5.1.3** Compare and contrast entry criteria and exit criteria (K2)
>
> **FL-5.1.4** Use estimation techniques to calculate the required test effort (K3)
>
> **FL-5.1.5** Apply test case prioritization (K3)
>
> **FL-5.1.6** Recall the concepts of the test pyramid (K1)
>
> **FL-5.1.7** Summarize the testing quadrants and their relationship with test levels and test types (K2)

In this section, we'll talk about a complicated pair of test topics: plans and estimates. Plans and estimates depend on a number of factors, including the level, targets and objectives of the testing we are setting out to do. Writing a plan and preparing an estimate tend to happen concurrently and ideally during the planning period for the overall project (called release planning in Agile development), though we must be ready to revise them as the project proceeds and we gain more information (which happens as part of iteration planning in Agile development).

Let's look closely at how to prepare a test plan, examining issues related to planning for a project, for a test level, for a specific test type and for test execution. We'll discuss ways to establish adequate entry and exit criteria for testing. In addition, we'll look at various test tasks that need estimating. We'll examine typical factors that influence the effort related to testing and see different estimation techniques.

Look out for the Glossary terms **entry criteria**, **exit criteria**, **test approach**, **test plan**, **test planning**, **test pyramid** and **testing quadrants** in this section.

5.1.1 The purpose and content of a test plan

While people tend to have different definitions of what goes in a **test plan**, for us a test plan is the project plan for the testing work to be done. It is not a test design specification, a collection of test cases or a set of test procedures; in fact, most of our test plans do not address that level of detail.

> **Test plan**
> Documentation describing the test objectives to be achieved and the means and the schedule for achieving them, organized to coordinate testing activities.

The syllabus talks about four main attributes of a test plan:

- The test plan should document the who, how, what, when, and where of testing, and explain how those activities relate to defined or implicit test objectives.

- The test plan should be aligned, in a transparent and traceable way, with any standards, criteria, regulations, internal templates, or other rules about organizational documentation.

- The test plan should be accessible to, and useful for, both testing participants and non-testing stakeholders.

- In addition to alignment with the test objectives, in organizations where a defined test policy and test strategy exist, the test plan should adhere—again in a transparent and traceable way—to the test policy and test strategy, except where there is a need to deviate from policy or strategy, in which case the deviation and the reason for it should be explained.

That's all true and useful, but let's also step back and ask ourselves, 'Why do we write test plans?' We have three main reasons: to guide our thinking, to communicate to others and to help manage future changes.

First, writing a test plan guides our thinking. We find that if we can explain something in words, we understand it. If we cannot explain it, there's a good chance we don't understand. Writing a test plan forces us to confront the challenges that await us and focus our thinking on important topics. In Chapter 2 of Fred Brooks' brilliant and essential book on software engineering management, *The Mythical Man-Month*, he explains the importance of careful estimation and planning for testing as follows:

> *Failure to allow enough time for system test, in particular, is peculiarly disastrous. Since the delay comes at the end of the schedule, no one is aware of schedule trouble until almost the delivery date [and] delay at this point has unusually severe ... financial repercussions. The project is fully staffed, and cost-per-day is maximum [as are the associated opportunity costs]. It is therefore very important to allow enough system test time in the original schedule.*

Brooks [1995]

This is particularly applicable to sequential development life cycles (as many were back in 1995). Agile development was designed to address some of these problems, but it does not obviate the need for system testing, so the advice is still relevant.

We find that using a template when writing test plans helps us remember the important challenges. You can use the template and examples shown in ISO/IEC/IEEE 29119-3 [2021], use someone else's template or create your own template over time.

The ISTQB Foundation Syllabus says that the test plan should cover:

- The context in which the testing occurs, including what is in and out of scope for testing, the objectives testing serves, and what the tests should be designed to cover (i.e. the test basis).

- Assumptions about, and constraints on, the testing to be done.

- Who the stakeholders are, what their roles and responsibilities are, how they relate to testing, and any hiring and training required to fill staffing or skills gaps.

- How we intend to communicate about testing, including any templates and dashboards to be used, along with the channels used to send those communications and the frequency of communication in each such channel (see also below).

- The project risks (at least those that affect testing) and the product risks (those that will—and won't—be addressed by testing), either explicitly or by reference to one or more risk registers.

Test approach
The manner of implementing testing tasks.

- The way that the test strategy will be implemented on the project (i.e. the **test approach**), including an explanation of any deviations from the organizational test strategy (or the test policy), discussion of the test levels, types, and techniques to be employed, the deliverables and metrics to be created, and what test data and environments are needed.

- The test entry and exit criteria (or definition of ready and definition of done on Agile projects) and how those (and the plan as a whole) are aligned with project-wide and/or organization-wide established criteria.

- The budget required and the planned schedule.

That said, the specific content that is necessary to serve the needs of the testers and the team on a project should drive the decisions about what to include, unless some regulatory requirements specify particular content.

Test planning   The activity of establishing or updating a test plan.

Second, the **test planning** process and the plan itself serve as vehicles for communicating with other members of the project team, testers, peers, managers and other stakeholders. This communication allows the test plan to influence the project team and allows the project team to influence the test plan. This is especially important in the areas of organization-wide testing policies and motivations; test scope, objectives and critical areas to test; project and product risks, resource considerations and constraints; and the testability of the item under test.

You can accomplish this communication through circulation of one or two test plan drafts and through review meetings. Such a draft may include many notes such as the following examples:

[To Be Determined: Jennifer: Please tell me what the plan is for releasing the test items into the test lab for each cycle of system test execution?]

[Dave – please let me know which version of the test tool will be used for the regression tests of the previous increments.]

As you document the answers to these kinds of questions, the test plan becomes a record of previous discussions and agreements between the testers and the rest of the project team. Thus, test planning is an ongoing activity performed throughout the product's life cycle (and beyond into maintenance).

Third, the test plan helps us manage change. During early stages of the project, as we gather more information, we revise our plans. As the project evolves and situations change, we adapt our plans. Written test plans give us a baseline against which to measure such revisions and changes. Furthermore, updating the plan at major milestones helps keep testing aligned with project needs. As we run the tests, we make final adjustments to our plans based on the results. You might not have the time—or the energy—to update your test plans every time a variance occurs, as some projects can be quite dynamic. A simple approach is described in Black [2009, chapter 6], for documenting variances from the test plan. You can implement it using a database or spreadsheet. You can include these change records in a periodic test plan update, as part of a test status report, or as part of an end of project test summary.

We have found that it is better to write multiple test plans in some situations. For example, when we manage both integration and system test levels, those two test execution periods may occur at different points in time and will have different objectives. For some systems projects, a hardware test plan and a software test plan will address different techniques and tools, as well as different audiences. However, since there might be overlap between these test plans, a master test plan that addresses the common elements can reduce the amount of redundant documentation.

What to do with your brain while planning tests

Writing a good test plan is easier than writing a novel, but both tasks require an organized approach and careful thought. In fact, since a good test plan is kept short and focused, unlike some novels, some might argue that it is harder to write a good test plan. Let's look at some of the planning tasks you need to carry out.

At a high level, you need to consider the purpose served by the testing work. In terms of the overall organizational needs, this purpose is referred to variously as the test team's mission or the organization's testing policy. A test plan is influenced by many factors; as well as test policy and the organization's test strategy, the development life cycle and methods being used for development, and the availability of resources. In terms of the specific project, understanding the purpose of testing means knowing the answers to questions such as:

- What is in scope and what is out of scope for this testing effort?
- What are the test objectives?
- What are the important project and product risks? (More on risks in Section 5.2.)
- What is the overall approach of testing in this project?
- How will test activities be integrated and coordinated into the software life cycle activities?
- How do we decide what to test, what people and other resources are needed to perform test activities, and how test activities will be carried out?
- What constraints affect testing (for example budget limitations, hard deadlines, etc.)?
- What is most critical for this product and project?

- Which aspects of the product are more (or less) testable?
- What should be the overall test execution schedule and how should we decide the order in which to do test analysis, test design, implementation, execution and evaluation of specific tests, either on specific dates or in the context of an iteration?
- What metrics will be used for test monitoring and control and how will they be gathered and analyzed?
- What is the budget for all test activities?
- What should be the level of detail and structure for test documentation?

In addition, you need to decide how to split the testing work into various levels, as discussed in Chapter 2 (for example, component, integration, system and acceptance). If that decision has already been made, you need to decide how to best fit your testing work in the level you are responsible for with the testing work done in those other test levels. During the analysis and design of tests, you will want to reduce gaps and overlap between levels and, during test execution, you will want to coordinate between the levels. Such details dealing with inter-level coordination are often addressed in the master test plan.

In addition to integrating and coordinating between test levels, you should also plan to integrate and coordinate all the testing work to be done with the rest of the project. For example, what items must be acquired for the testing? Are there ongoing supply issues, such as with imitation bank notes (that is, simulated bank notes) for a financial application such as an ATM? When will the developers complete work on the component or system under test? What operations support is required for the test environment? What kind of information must be delivered to the maintenance team at the end of testing?

Moving down into the details, what makes a plan a plan (rather than a statement of principles, a laundry list of good ideas or a collection of suggestions) is that the author specifies in it who will do what when and (at least in a general way) how. Resources are required to carry out the work. There are often hard decisions that require careful consideration and building a consensus across the team, including with the project manager.

The entire testing process, from planning through to completion, produces information, some of which you will need to document. How precisely should testers write the test designs, cases and procedures? How much should they leave to the judgement of the tester during test execution, and what are the reproducibility issues associated with this decision? What kinds of templates can testers use for the various documents they will produce? How do those documents relate to one another? If you intend to use tools for tasks such as test design or execution, as discussed in Chapter 6, you will need to understand how the models or automated tests will integrate with manual testing, and plan who will be responsible for automation design, implementation and support. There may be a separate test automation plan, but this needs to be coordinated with other test plans.

Some information you will need to gather in the form of raw data and then distil. What metrics to do you intend to use to monitor, control and manage the testing? Which of those metrics, and perhaps other metrics, will you use to report your results? We'll look more closely at possible answers to those questions in Section 5.3, but a good test plan provides answers early in the project.

Thinking about the test approach

A **test strategy** is the general way in which testing will happen within each of the levels of testing, independent of project, across the organization. While the ISTQB Glossary definition is a bit inscrutable and cryptic, the test approach is basically the implementation of the test strategy on a specific project. Since the test approach is specific to a project, you should define and document the approach in the test plans, refining and providing further detail in the test designs.

Deciding on the test approach involves careful consideration of the testing objectives, the project's goals and overall risk assessment. These decisions provide the starting point for planning the test process, for selecting the test design techniques and test types to be applied and for defining the entry and exit criteria. In your decision-making on the approach, you should take into account the project, product and organization context, issues related to risks, hazards and safety, the available resources, the team's level of skills, the technology involved, the nature of the system under test, considerations related to whether the system is custom-built or assembled from commercial off-the-shelf components (COTS), the organization's test objectives and any applicable regulations.

The choice of test approaches or strategies is one powerful factor in the success of the test effort and the accuracy of the test plans and estimates. Of course, having choices also means that you can make mistakes, so we'll go into more detail about how to pick the right test strategies shortly.

First, though, we will survey the major types of test strategies that are commonly found—this material is beyond what is in the current Syllabus and is included here for additional information which should be useful for you.[1]

- **Analytical:** In this strategy, tests are determined by analyzing some factor, such as requirements (or other test basis) or risk. For example, the risk-based strategy involves performing a risk analysis using project documents and stakeholder input, then planning, estimating, designing and prioritizing the tests based on risk. We will talk more about risk analysis later in this chapter. Another analytical test strategy is the requirements-based strategy, where an analysis of the requirements specification forms the basis for planning, estimating and designing tests. Analytical test strategies have in common the use of some formal or informal analytical technique, usually during the requirements and design stages of the project.

- **Model-based:** In this strategy, tests are designed based on some model of the test object. For example, you can build mathematical models for loading and response for e-commerce servers, and test based on that model. If the behaviour of the system under test conforms to that predicted by the model, the system is deemed to be working. Model-based test strategies have in common the creation or selection of some formal or informal model for critical system behaviours, usually during the requirements and design stages of the project. Examples also include business process models, state models, for example state transition diagrams, etc. or reliability grown models.

---

[1] This catalogue of testing strategies grew out of an email discussion between Rex Black, Ross Collard, Kathy Iberle and Cem Kaner. We thank them for their thought-provoking comments.

- **Methodical:** In this strategy, a pre-defined and fairly stable list of test conditions is used. For example, you might have a checklist that you have put together over the years that suggests the major areas that testing should cover, or you might follow an industry standard for software quality, such as ISO/IEC 25010 [2011], for your outline of major test areas. You then methodically design, implement and execute tests following this outline. Methodical test strategies have in common the adherence to a pre-planned, systematized approach. This may have been developed in-house, assembled from various concepts developed in-house and gathered from outside, or adapted significantly from outside ideas, and may have an early or late point of involvement for testing. Some examples include working through a list (that is, taxonomy) of typical defects, or working through a list of desired quality characteristics, such as company-wide look-and-feel standards for websites and mobile apps.

- **Process- or standard-compliant:** In this strategy, an external standard or set of rules is used to analyze, design and implement tests. For example, you might adopt the standard ISO/IEC/IEEE 29119-3 [2013] for your testing, or you may use TMMi (Test Maturity Model integration, www.tmmi.org) to assess your current testing and improve it. More information about TMMi can be found in van Veenendaal and Wells [2012]. Alternatively, process- or standard-compliant strategies have in common reliance upon an externally developed approach to testing, often with little (if any) customization. They may have an early or late point of involvement for testing. For example, some organizations need to conform to safety-critical or financial regulations which may include processes for identifying tests in a rigorous way from the test basis.

- **Directed (or consultative):** In this strategy, stakeholders or experts (technology or business domain experts) may direct the testing according to their advice and guidance. For example, you might ask the users or developers of the system to tell you what to test or even rely on them to do the testing. Consultative or directed strategies have in common the reliance on a group of non-testers to guide or perform the testing effort, and typically emphasize the later stages of testing simply due to the lack of recognition of the value of early testing. This strategy may be helpful when the developing organization is a new start-up without a lot of testing knowledge or expertise. For example, an outside expert may advise on security testing.

- **Regression-averse:** In this strategy, the most important factor is to ensure that the system's behaviour does not deteriorate or get worse when it is changed and enhanced. To protect existing functionality, automated regression tests would be extensively used, as well as standard test suites and reuse of existing tests and test data. For example, you might try to automate all the tests of system functionality so that, whenever anything changes, you can re-run every test to ensure nothing has broken. Regression-averse strategies have in common a set of procedures—usually automated—that allow them to detect regression defects. A regression-averse strategy may involve automating functional tests prior to release of the function, in which case it requires early testing, but sometimes the testing is almost entirely focused on testing functions that have already been released, which is in some sense a form of post-release test involvement.

- **Reactive (or dynamic):** In this strategy, the tests react and evolve based on what is found while test execution occurs, rather than being designed and implemented before test execution starts. For example, you might create a lightweight set of testing guidelines that focus on rapid adaptation or known weaknesses in software. Reactive strategies, using exploratory testing, have in common concentrating on finding as many defects as possible during test execution and adapting to the realities of the system under test as it is when delivered, and they typically emphasize the later stages of testing.

Some of these strategies are more preventive, others more dynamic. For example, analytical test strategies involve up-front analysis of the test basis and tend to identify problems in the test basis prior to test execution. This allows the early, cheap removal of defects. That is a strength of preventive approaches.

Reactive test strategies focus on the test execution period. Such strategies allow the location of defects and defect clusters that might have been hard to anticipate until you have the actual system in front of you. That is a strength of reactive approaches.

Rather than see the choice of strategies, particularly the preventive or reactive strategies, as an either/or situation, we'll let you in on the worst kept secret of testing (and many other disciplines). There is no one best way. We suggest that you adopt whatever test strategies make the most sense for your particular test approach, and feel free to borrow and blend.

How do you know which strategies to pick or blend for the best chance of success? There are many factors to consider, but let's highlight a few of the most important:

- **Risks:** Testing is about risk management, so consider the risks and the level of risk. For a well-established application that is evolving slowly, regression is an important risk, so regression-averse strategies make sense. For a new application, a risk analysis may reveal different risks if you pick a risk-based analytical strategy.

- **Skills:** Strategies must not only be chosen, they must also be executed. So, you have to consider the skills that your testers possess and lack. A standard-compliant strategy is a smart choice when you lack the time and skills in your team to create your own approach.

- **Objectives:** Testing must satisfy the needs of stakeholders to be successful. If the objective is to find as many defects as possible with a minimal amount of up-front time and effort invested—for example, at a typical independent test lab—then a dynamic or reactive strategy makes sense.

- **Regulations:** Sometimes you must satisfy not only stakeholders, but also regulators. In this case, you may need to devise a methodical test strategy that satisfies these regulators that you have met all their requirements.

- **Product:** Some products such as weapons systems and contract development software tend to have well-specified requirements. This leads to synergy with a requirements-based analytical strategy.

- **Business:** Business considerations and business continuity are often important. If you can use a legacy system as a model for a new system, you can use a model-based strategy.

We mentioned above that a good team can sometimes triumph over a situation where materials, process and delaying factors are ranged against its success. However, talented execution of an unwise strategy is the equivalent of going very fast down a motorway in the wrong direction. Therefore, you must make smart choices in terms of testing strategies. Furthermore, you must choose testing strategies with an eye toward the factors mentioned earlier, the schedule, budget and feature constraints of the project and the realities of the organization and its politics.

5.1.2 Tester's contribution to iteration and release planning

If you are a tester in an Agile team, you are likely to be involved in planning activities that might have been done primarily by test managers in your previous projects. So, test planning is something that you'll need to learn more about if you don't have a background in or experience doing test planning.

Even if you do have experience, you still have something to learn here, because there's a difference in the way that the planning happens. First, there's release planning. Release planning happens prior to the first iteration. It's about defining what we are going to build in this release. It's an initial cut at the product backlog, which is sometimes called the release backlog.

Depending on the flavour of Agile you're following, and the tools used to support it, the product backlog often consists of user stories and epics, where an epic is a sequence of user stories that are dependent on or related to each other. Epics are typically built over a sequence of iterations. Again, depending on the flavour of Agile and the tools in use, user stories can contain tasks, and in some cases those tasks will be test-related tasks associated with the specific feature or functionality being built via the user story.

Release planning

Your test plan, including your test approach, needs to be defined during release planning. Your test plan needs to define the general way in which you're going to carry out your testing activities on an iteration-by-iteration basis, independent of the content of the iteration.

That last bit—independent of the content of the iteration—is really important. People are often used to writing test plans that focus on the content of the testing effort. In an Agile project that is not something that can be predefined with great precision, because things can be added to and deleted from the product backlog at any point. So, you don't know the overall content of every iteration and you don't even know for sure in which iteration a particular user story is going be defined, built and tested.

However, you can set out some general rules about how you do things from one iteration to the next. That's what the test plan should do. There should also be a high-level risk analysis, which is something we'll talk more about later in this chapter. There should also be a high-level estimation of effort for the user stories that are defined in the product backlog, including the test effort. The release plan is subject to change, just like the product backlog is. So, release planning is something that should be kept flexible.

Keep in mind, as you do your release planning, that there might be testing tasks that cannot be done within each iteration. The test plan should define these as well. Examples include security testing on the whole product or some special load testing.

Now, one of the things you definitely need to do as part of release planning is to define your test basis and your test oracles. Remember, your test basis consists of the things that you base your tests on. You should not assume that covering the user stories and their acceptance criteria is sufficient. That's just the Agile version of a classic testing mistake, which is using a purely requirements-based testing strategy. You need to expand your thinking about your test basis.

Your test oracles, remember, are the things that you're going to refer to in order to define the expected results of your tests. This can include the user story as one of the elements, but also think about things like legacy systems and competitor systems. Think expansively about what your test oracles are.

As a tester in an Agile team, you don't just search for your test basis and test oracles: you help to define them. You should participate in the writing of the user stories that are going into the product backlog, especially the acceptance criteria for those user stories. (By the way, this is the time to leverage the whole team approach we talk about in Chapter 1.) You want to ensure that the user stories are testable and that the acceptance criteria are ones that you can actually evaluate via a test case.

Iteration planning

Iteration planning, naturally enough, happens at the beginning of each iteration. We select from the product backlog the user stories we want in our sprint backlog—or iteration backlog if you prefer. The selection is done based on priority as defined by the product owner, but also on other considerations. For example, if a series of user stories are part of an epic, there might be a natural sequence in which to build them. The user stories that are selected often need to be elaborated and made more specific.

You might have to adjust your test plan as part of iteration planning. These adjustments should be small. If you're having to significantly update your test plan at the beginning of each iteration, you might not be doing it right. Perhaps your test plan's not content independent.

Let's explain what we mean by 'content independent' in this context. If you write a test plan that makes assumptions about specific content in an Agile project, as the product backlog and product increment change, those assumptions may be violated. In Agile, the test plan must address how testing will be done, independent of changes that might occur with regard to what is to be tested. Such a test plan can be considered content independent.

You will need to do risk analysis as part of iteration planning. We'll discuss how to do that later in this chapter. You'll need to estimate the test effort associated with each user story—more on that later in this section—and that needs to be reconciled with the team's actual velocity (i.e. the team's demonstrated ability in terms of the number of user stories that they can complete per iteration).

If the user story is not clear in terms of what it means, the product owner can be involved in trying to get the user story clarified. Theoretically, if clarification is not available, then the user story should be returned to the product backlog. Presumably, the next time that user story is selected, the problems with it will be resolved. Personally, we find that option somewhat problematic. Basically, this is like saying to the product owner, 'Well, since you couldn't be bothered to resolve my questions about this particular feature that you want, I can't be bothered to build and test it right now.' Not a real collaborative kind of behaviour, so hopefully that doesn't happen very often.

At the end of the iteration planning, we should have a set of user stories and tasks defined (possibly on a task board), including test-related tasks. Each of those has clear ownership, so we know who's responsible for doing what.

In one of the most popular forms of Agile, Scrum, the theory is that the content of a sprint is not supposed to change once iteration planning is done. In practice, we do see that happen when consulting with organizations using Scrum. In some cases, it includes adding new user stories, and, when that happens, it is disruptive, especially to the testers. Even when it involves removing user stories, though, this can be disruptive to the team and to the testers.

As testers, there are some things you specifically need to do in iteration planning. Certainly, performing the product risk analysis is something in which the tester is a primary participant. Testers should also evaluate the user stories with respect to testability and collaborate with the business and technical stakeholders to define the acceptance criteria. You'll need to figure out what test tasks are needed for the different user stories, create the tests that will be run against the user stories and review those tests with the business stakeholders and the developers.

In addition, you should work with the developers to determine which levels of testing need to be run. (We'll return to this topic of test levels in Agile projects when we get to the material on testing quadrants and the test pyramid a little later in this chapter.) You should discuss with the developers the functional and non-functional tests that are needed, being careful not to neglect non-functional testing. Ask questions like, 'What about performance? What about reliability? What about security? What about usability?'

Planning for automation is important in iteration planning. Automation is a challenge and an opportunity for many testers working in Agile worlds. Step up to that challenge and increase your technical skills, if need be, and be ready to participate in automation. During iteration planning, make sure you're ready to accommodate changes as they occur.

Finally, remain in close communication with the team throughout the iteration and planning process. Usually, this iteration planning is done within a day or maybe two days depending on the length of the sprint. You as a tester should be involved in the entire process. If you're not, then you're not being treated as a co-equal member of the whole team, and that's not Agile.

5.1.3 Entry criteria and exit criteria (definition of ready and definition of done)

Two important things to think about when determining the approach and planning the testing are: how do we know we are ready to start a given test activity, and how do we know we are finished (with whatever testing we are concerned with)? At what point can you safely start a particular test level? When are you confident that it is complete? The factors to consider in such decisions are called entry and exit criteria or in Agile development, definition of ready and definition of done.

Typical **entry criteria** include the following:

Entry criteria
(definition of ready)
The set of conditions for officially starting a defined task.

- Availability of testable requirements, user stories and/or models, for example when following a model-based testing strategy, that is, the test basis is available.
- Availability of test items that have met the exit criteria for any previous test levels.
- Availability of the test environment.
- Availability of necessary test tools and any other materials needed.
- Availability of testware, including test cases, test data and other necessary resources.

- Availability of budget, time and staff for testing tasks.
- Availability of the component or system to be tested, in a state where tests can be done, that is, availability of the test object.

Why are entry criteria important? If everything needed for testing to go ahead is in place before you start, the testing will go much more smoothly. The problems of getting started with the testing often turn out to be that something needed is not actually ready or in place. Then the testing process gets the blame for the delays. If entry criteria, which are the preconditions for testing, are enforced, or at least thought about beforehand, everything is more likely to go better. If not, you are increasing risk, introducing delays and additional costs, and making life more difficult for yourself.

Typical **exit criteria** include the following:

- Tests: the number planned, prepared, run, passed, failed, blocked, skipped, etc. are acceptable (possibly taking into account both dynamic and static tests), and, in some cases, whether regression tests have been automated.
- Coverage: the extent to which the test basis (for example, requirements, user stories, acceptance criteria), risk, functionality, supported configurations, and the software code have been tested (that is, achieved a defined level of coverage) —or have not.
- Defects: the number known to be present, the arrival rate, the number estimated to remain, the number resolved and the number of unresolved defects are within an agreed limit.
- Quality: the status of the important quality characteristics for the system, for example reliability, performance efficiency, usability, security and other relevant quality characteristics are adequate.
- Money: the cost of finding the next defect in the current level of testing compared to the cost of finding it in the next level of testing (or in production).
- Schedule: the project schedule implications of starting or ending testing.
- Risk: the undesirable outcomes that could result from shipping too early (such as latent defects or untested areas), or too late (such as loss of market share).

> **Exit criteria**
> (completion criteria, test completion criteria, definition of done) The set of conditions for officially completing a defined task.

In the syllabus, the items related to tests, coverage and defects are referred to as completion criteria and thoroughness measures.

When writing exit criteria, we try to remember that a successful project or iteration is a balance of quality, budget, schedule and feature considerations. This is important in each sprint and is important in other development life cycles when exit criteria are applied at the end of the project.

Why are exit criteria important? Knowing what your goals are is important in any human endeavour. If we do not think about 'How will we know we are done' beforehand, then we may stop before we have done enough testing (increasing risk) or we might keep testing when we have done enough (not likely but possible, and this would be wasteful).

In practice, testing is often stopped rather than finished, because time pressure often prevails over other considerations. But if (or rather when) this happens, if you do have documented exit criteria, you can make the risks more visible to stakeholders and managers by showing what has not yet been completed in the testing. You may not win the argument for doing more testing at the time, but you will have good information to explain next time why the exit criteria are important, especially when/if there are undesirable consequences to not meeting them this time. We also want a 'Stop testing now!' decision to be made with knowledge of all the factors, not just basing that decision on a calendar date.

5.1.4 Estimation techniques

Test estimation
An approximation related to various aspects of testing.

In this section, we'll look at **test estimation**, first describing how to go about estimating testing, what it will involve and what it can cost. We will then look at factors that influence the test effort, many of which are significant for testing even though they are not part of what we are estimating when we estimate testing. Finally, we'll cover four specific, complementary techniques for taking all the information we cover in this subsection and using it to come up with a solid estimate.

Estimating what testing will involve and what it will cost

The testing work to be done can often be seen as a subproject within the larger project. We can adapt fundamental techniques of estimation for testing. We could start with a work-breakdown structure that identifies the stages, activities and tasks.

Starting at the highest level, we can break down a testing project into major activities using the test process identified in the ISTQB Syllabus (and described in Chapter 1). Within each activity, we identify tasks and perhaps subtasks. To identify the activities and tasks, we work both forward and backward. When we say we work forward, we mean that we start with the planning activities and then move forward in time step-by-step, asking 'Now, what comes next?'

Working backward means that we consider the risks that we identified during risk analysis (which we'll discuss in Section 5.2). For those risks which you intend to address through testing, ask yourself, 'What activities and tasks are required in each stage to carry out this testing?' Let's look at an example of how you might work backwards.

Suppose that you have identified performance as a major area of risk for your product. Performance testing is an activity in test execution. You now estimate the tasks involved with running a performance test, how long those tasks will take and how many times you will need to run the performance tests.

Now, those tests did not just appear out of thin air: someone had to develop them. Performance test development entails activities in test analysis, design and implementation. You now estimate the tasks involved in developing a performance test, such as writing test scripts and creating test data.

Typically, performance tests need to be run in a special test environment that is designed to look like the production or field environment, at least in those ways which would affect response time and resource utilization. Performance test environment acquisition and configuration is an activity in the test implementation activity. You now estimate tasks involved in acquiring and configuring such a test environment, such as simulating performance based on the production environment design to look for potential bottlenecks, getting the right hardware, software and tools and setting up that hardware, software and tools. Performance tests need special tools to generate load and check response. The acquisition and implementation of such tools also need to be planned.

It may be possible to use virtualization for your test environment and performance testing tools; this seems to be an attractive idea, since it should save money as you do not have to acquire your own environment or tools. However, the use of virtual environments and using the tools in those environments still needs careful planning.

Not everyone knows how to use performance testing tools or to design performance tests. Performance testing training or staffing is a task in the test planning activity. Depending on the approach you intend to take, you now estimate the time required to identify and hire a performance test professional or to train one or more people in your organization to do the job.

Finally, in many cases a detailed test plan is written for performance testing, due to its differences from other test types. Performance testing planning is a task in test planning. You now estimate the time required to draft, review and finalize a performance test plan.

When you are creating your work-breakdown structure, remember that you will want to use it for both estimation (at the beginning) and monitoring and control (as the project continues). To improve the accuracy of the estimate and enable more precise control, make sure that you subdivide the work finely enough. This means that tasks should be short in duration, say one to three days. If they are much longer—say two weeks—then you run the risk that long and complex subtasks are hiding within the larger task, only to be discovered later. This can lead to nasty surprises during the project.

Factors that affect the test effort

Testing is a complex endeavour on many projects and a variety of factors can influence it. When creating test plans and estimating the testing effort and schedule, you must keep these factors in mind, or your plans and estimates will deceive you at the beginning of the project and betray you at the middle or end.

The test strategies or approaches you pick will have a major influence on the testing effort. In this section, let's look at factors related to the product, the development process, the people involved and the results of testing.

Product characteristics

The characteristics of the product that we are testing have a major impact on how we will test it:

- The risks associated with the product. A high-risk product needing a lot of testing will suffer much more severe impacts if testing is not estimated correctly, especially if it is underestimated.

- The quality of the test basis. We want sufficient product documentation so that the testers can figure out what the system is, how it is supposed to work and what correct behaviour looks like. In other words, adequate and high-quality information about the test basis will help us do a better, more efficient job of defining the tests.

- The size of the product. A larger product leads to increases in the size of the project and the project team. This will also increase the difficulty of predicting and managing the projects and the team. This leads to the disproportionate rate of collapse of large projects.

- The requirements for quality characteristics such as usability, reliability, security, performance, etc. also influences the testing effort. These test types can be expensive and time-consuming.

- The complexity of the product domain. Examples of complexity considerations include:
  - The difficulty of comprehending and correctly handling the problem the system is being built to solve, for example avionics and oil exploration software.

- The use of innovative technologies, especially those long on hyperbole and short on proven track records.
- The need for intricate and perhaps multiple test configurations, especially when these rely on the timely arrival of scarce software, hardware and other supplies.
- The prevalence of stringent security rules, strictly regimented processes or other regulations.

- Requirements for legal and regulatory compliance. If you are working in a regulated industry, the time and effort taken to meet those regulatory requirements can be significant.

Development process characteristics

The life cycle and development process in use has an impact on how the test effort is spent. Testing is quite different in Agile development compared to a sequential development life cycle, and other aspects of development also influence testing:

- The stability and maturity of the organization. Mature organizations tend to have better requirements, architecture and unit tests, thus saving test effort later in the life cycle.

- The development life cycle model in use. The life cycle model itself is an influential process factor, as the V-model tends to be more fragile in the face of late change, while incremental models, such as Agile development, tend to have high regression testing costs.

- The test approach. Choosing the right approach is important for good testing. With a less than ideal approach, testing will take longer and take more effort than is necessary.

- The tools used. Tools are supposed to increase efficiency and reduce time spent on some tasks, so test tools, especially those that reduce the effort associated with test execution, which is on the critical path for release, should decrease execution time for tests. Of course, other factors about automation may be far more significant, as we will see in Chapter 6. On the development side, debugging tools and a dedicated debugging environment (as opposed to debugging in the test environment) also reduce the time required to complete testing.

- The test process. A test process that is well understood, with testers trained to perform the activities and tasks they need to do in the most effective and efficient way, is the optimum. Test process maturity is another factor, since more mature testing will be more efficient and effective.

- The geographical distribution of the team, especially if the team crosses time zones (as many outsourcing efforts do).

- The required level of detail for test documentation. While good project documentation is a positive factor, it is also true that having to produce detailed documentation, such as meticulously specified test cases, results in delays. During test execution, having to maintain such detailed documentation requires lots of effort, as does working with fragile test data that must be maintained or restored frequently during testing.

- Time pressure. This is another factor to be considered. Pressure should not be an excuse to take unwarranted risks. However, it is a reason to make careful, considered decisions and to plan and re-plan intelligently throughout the process, which is another hallmark of mature processes.

People characteristics

People execute the process, and people factors are as important or more important than any other. Indeed, even when many troubling things are true about a project, an excellent team can often make good things happen on the project and in testing. Important people factors include:

- The skills and experience of the people involved. The skills of individuals and the team as a whole are important, as well as the alignment of those skills with the project's needs. Domain knowledge is likely to be more relevant when the problem being solved is complex, thus requiring specific knowledge on the part of the tester to operate the software and to determine correct versus incorrect behaviour.
- Team cohesion and leadership. Since a project team is a team, solid relationships, reliable execution of agreed-upon commitments and responsibilities and a determination to work together toward a common goal are important. This is especially important for testing, where so much of what we test, use and produce either comes from, relies upon or goes to people outside the testing group. Because of the importance of trusting relationships and the lengthy learning curves involved in software and system engineering, the stability of the project team is an important people factor, too.

Test results

The test results themselves are important in the total amount of test effort during test execution:

- The number and severity of defects found. Wouldn't testing be easy if we never found any defects? Initial tests would just run with no problems, and there would be no need for any tests to be repeated. However, in the real world, the more defects there are and the more severe those defects, the greater the impact on the testing and the test estimates.
- The amount of rework required. Delivery of good quality software at the start of test execution and quick, solid defect fixes during test execution prevents delays in the test execution process. A defect, once identified, should not have to go through multiple cycles of fix/retest/re-open, at least not if the initial estimate is going to be held to. Good design of the test object should make changes easier. For example, good modular design may have a low-level function called by a number of higher-level functions. If a change is made in the low-level function, the functions calling it should work without being changed (if correctly designed).

Note that, with good historical data from similar past projects, the number of defects found, their severity and the associated rework of fixing defects can be predicted, often with surprising accuracy.

You probably noticed from this list that we included a number of factors outside the scope and control of the testers. Indeed, events that occur before or after testing can bring these factors about. For this reason, it is important that testers, especially test managers, be attuned to the overall context in which they operate. Some of these contextual factors result in specific project risks for testing, which should be addressed in the test plan. Project risks are discussed in more detail in Section 5.2.

Four test estimation techniques

There are four techniques for estimation covered by the ISTQB Foundation Syllabus. Two involve consulting the people who will do the work and other people with expertise on the tasks to be done (expert-based). The other two involve analyzing metrics from past projects and from industry data (metrics-based). Let's look at each in turn.

Expert-based: Wideband Delphi and three-point estimation

Asking the individual contributors and experts often involves working with experienced staff members to develop a work-breakdown structure for the project, or, in an Agile project, to assign effort estimates to user stories. (We'll discuss Agile estimation in more detail in the next subsection.) This includes working together to understand, for each task, the effort, duration, dependencies and resource requirements. The idea is to draw on the collective wisdom of the team to create your test estimate. Using project management software or a whiteboard and sticky notes, you and the team can then predict the testing end date and major milestones in a sequential project or determine how many user stories will actually fit in the iteration. This technique is often called bottom-up estimation because you start at the lowest level of the hierarchical breakdown, either in the work-breakdown structure (the task) or the tasks associated with individual user stories, and let the duration, effort, dependencies and resources for each task add up across all the tasks. These are expert-based techniques.

To put more formality around these expert-based techniques, we sometimes see either the Wideband Delphi approach, the three-point approach, or both combined. In Wideband Delphi, the experts participating in the estimation are asked to estimate each task in isolation, without consulting with others. This is done to avoid what is called anchoring, which is where, in a discussion, the first idea put forth tends to anchor the subsequent discussion. We want each person thinking freely in their estimation. Once the estimates are collected, we check to see if we have a reasonably high degree of agreement (i.e. some minor differences are often allowed). If we do not, then the high and low estimator give their justifications for their estimates, and the estimators again—separately—repeat estimation. This process can iterate, though usually a rule is in place to limit the number of iterations to avoid giving one recalcitrant expert an effective heckler's veto by refusing to compromise.

Three-point estimation expands on the idea of having experts estimate by asking them to give not one, but three estimates:

- The most optimistic estimation, which is assigned to a.
- The most likely estimation, which is assigned to m.
- The most pessimistic estimation, which is assigned to b.

The final estimate, assigned to E, is a weighted arithmetic mean of a, m, and b. While there are variations in how that mean is calculated, the syllabus cites:

$$E = \frac{(a + 4m + b)}{6}$$

The user of this formula also allows the calculation of a standard deviation, which the syllabus refers to as the measurement error, using the formula:

$$SD = \frac{(b - a)}{6}$$

To illustrate the concept, the syllabus gives the following example. If an estimator assigns a = 6, m = 9, and b = 18, then the final estimation E is 10, since:

$$\frac{(6 + 4 \times 9 + 18)}{6} = 10$$

The standard deviation SD, or measurement error, is 2, since:

$$\frac{(18 - 6)}{6} = 2$$

Therefore, we would expect the task to take 10 person-hours, give or take two person-hours.

Metrics-based: estimation based on ratios and extrapolation

Analyzing metrics can be as simple or sophisticated as you make it. The simplest approach is to ask, 'How many testers do we typically have per developer on a project?' A somewhat more reliable approach involves classifying the project in terms of size (small, medium or large) and complexity (simple, moderate or complex) and then seeing on average how long projects of a particular size and complexity combination have taken in the past. Another simple and reliable approach we have used is to look at the average effort per test case in similar past projects and to use the estimated number of test cases to estimate the total effort. Sophisticated approaches involve building mathematical models in a spreadsheet or other tool that look at historical or industry averages for certain key parameters—number of tests run by a tester per day, number of defects found by a tester per day, etc.—and then plugging in those parameters to predict duration and effort for key tasks or activities on your project. The tester-to-developer ratio is an example of a top-down estimation technique, in that the entire estimate is derived at the project level, while the parametric technique is bottom-up, at least when it is used to estimate individual tasks or activities.

Defect removal models are another example of metrics-based techniques. Data is gathered from previous projects about the number of defects and the time to remove them; this provides a basis for future similar projects.

The Foundation syllabus specifically talks about two types of metrics-based estimation. The first is the use of effort or staff ratios. Ideally, these are ratios derived from previous, similar projects, rather than relying on industry ratios, due to the extreme variation seen in such industry numbers and the fact that a lack of context around those industry ratios makes them hard to adjust for internal realities. The syllabus gives an example where, across a number of previous projects the development-to-test effort ratio was 3:2, so, if the development effort is estimated at 600 person-days, the test effort should be estimated at 400 person-days. This technique can be used on sequential or Agile projects, albeit on Agile projects it would tend to apply more during release planning than iteration planning.

The second type of metrics-based estimation, extrapolation, is particularly suited for Agile projects, including during iteration planning. We begin to take measurements from the start of the current project, and, as we continue along in the project, we start to use that data to estimate the effort required for the remaining work through extrapolation. The syllabus gives the example of estimating the test effort for the next iteration by using the averaged test effort from the last three iterations.

In Agile development, planning poker is an example of the expert-based technique; team members estimate based on their own experience of the effort needed to deliver and test a feature. Burndown charts are an example of the metrics-based technique, since the effort being spent is captured and reported and used to feed into the team's velocity to determine the amount of work the team can do in the next iteration. This is actual monitoring of the current sprint to use the results to help predict and estimate the effort needed for the following sprint. We'll discuss Agile estimation further in the next subsection.

How to best approach this mix of techniques for estimation? We prefer to start by drawing on the team's wisdom to create the work-breakdown structure and a detailed bottom-up estimate. We then apply metrics, models and rules of thumb to check and adjust the estimate bottom-up and top-down using past history. This approach tends to create an estimate that is both more accurate and more defensible than either technique by itself.

Even the best estimate must be negotiated with management. Negotiating sessions exhibit amazing variety, depending on the people involved. However, there are some classic negotiating positions. It is not unusual for the tester or test manager to try to sell the management team on the value added by the testing or to alert management to the potential problems that would result from not testing enough. It is not unusual for management to look for smart ways to accelerate the schedule or to press for equivalent coverage in less time or with fewer resources. In between these positions, you and your colleagues can reach compromise, if the parties are willing. Our experience has been that successful negotiations about estimates are those where the focus is less on winning and losing and more about figuring out how best to balance competing pressures in the realms of quality, schedule, budget and features.

Estimation in Agile projects

In Agile, as in sequential life cycles, product risk analysis, which we'll discuss later in this chapter, should precede and inform test estimation. However, in Agile, the two efforts might be part of the same team session, incorporated into the larger process of iteration planning. As mentioned earlier, iteration planning starts with the selection of the user stories from the release backlog. Once the user stories have been selected, the risk analysis can begin. Once the risk analysis is concluded, you now have an idea of how much testing effort is associated with each user story, as well as the testing effort associated with activities that may not be user story specific, such as performance testing or security testing. So, in aggregate, we know the amount of test effort required. This becomes an input to the estimation process.

Different Agile organizations estimate in different ways. Some of our clients that use Scrum estimate with some variation of planning poker. Planning poker is basically a variant of the Wideband Delphi technique discussed earlier.

In planning poker, the team sizes the user stories and other tasks in terms of the amount of effort required. Some of our clients size in person-hours. Some use what are called story points, where the story points typically follow some variation of the

Fibonacci series. We've also heard of people using tee-shirt sizes, such as small, medium, large, and extra large. If person-hours or story points are used, a burndown chart can be created.

The entire team should be involved in estimation. As a tester, you are a co-equal member of this process. You must be careful not to miss any important tasks. For example, there might be test-enabling tasks for test data creation, test environment configuration, network configuration, or so forth. Such enabling tasks are required to test a particular user story. So, you should think very carefully about such tasks. Of course, you should encourage other people on the team to think carefully about any important issues that are related to their particular tasks.

Estimation should also use metrics based techniques in Agile projects, often using the concept of velocity. Velocity has to do with how many story points or person-hours of effort a team has actually accomplished in a given iteration. Velocity is often tracked using a burndown chart. When estimating during iteration planning, past velocity can be used to avoid overcommitting, as selection of a set of user stories whose total person-hours or story points exceeds past velocity would be a warning sign of taking on too much work.

As you use burndown charts, you'll be able to track your actual velocity, and use that to refine your ability to estimate for future iterations. If, in the course of a particular iteration, you find out that you've bitten off more than you can chew, then those user stories which cannot satisfy their definition of done within the context of the sprint would go back in the product backlog. As a practical matter, these user stories often go into the iteration backlog for the next sprint.

One of the worst practices that we've seen happens when groups over-commit in their estimates and they extend the duration of the iteration or sprint. You can make an argument for doing so by saying, 'Look, if we just have one more week, we can get these extra things done and we really need these done in the iteration.' However, you have to be careful with that argument as an organization because it can degrade the kind of discipline that you need to have to carry out Agile methods properly.

Another bad practice that we've seen happen—one that we consider worse than extending the iteration—is that some of our clients have their sprints end on a Friday, and they specifically reserve the following Saturday and Sunday to catch up. That leads to these Saturdays and Sundays being these arbitrary, overtime, crunch modes. The brunt of this crunch usually falls on the test team, as they have to come to work and finish testing the things that were not ready for testing on Friday.

Such practices burn out the testers and are antithetical to the idea of a sustainable workload. A better practice is to have the iterations end on a Tuesday or a Wednesday. That way, the dangling temptation of using the weekend to finish the iteration is reduced. This tendency to make the testers work weekends to meet arbitrary iteration end dates is a significant dysfunction that we have observed with a number of Agile teams. Over-commitment is a basic problem in software engineering. Simply adopting Agile methods won't change the underlying human elements that lead organizations to over-commit.

The learning objective to use estimation techniques is a K3 in the Syllabus, which means that you need to be able to produce an estimate. Refer to the exercise at the end of this chapter, and the mock exam in Chapter 7.

5.1.5 Test case prioritization

Part of test management is the management of tests. Once test cases and test procedures have been designed and implemented (some as automated tests), they may be assembled into test suites for convenience of running sets of tests together.

Ideally, the test cases would be run in priority order. Priority can be determined using one or more of the following techniques:

- Risk-based prioritization. In the next section, we'll talk about risk analysis. In risk-based prioritization, the order in which tests are run is determined by the level of risk associated with each test.

- Coverage-based prioritization. In this technique, some measure of coverage, often code coverage, is used to determine which tests should go first, based on a higher coverage number. This can be adjusted to look at additional coverage after a given set of tests are run, i.e. prioritizing the tests that will cover the most previously uncovered aspects of the system.

- Requirements-based prioritization. Similar to risk-based prioritization, this technique also relies on traceability, but in this case, it is between a requirement and the tests that cover it, rather than a risk and the tests that cover it. The priority given to the requirement—ideally based on a consensus of the stakeholders—is transferred to the related test cases.

However, in actuality, a schedule for the execution of the test suites must be based on a number of factors, including the priorities of the tests, but also technical or logical dependencies of tests or test suites and the type of tests, for example confirmation tests after defects have been fixed, or regression tests. These factors need to be balanced with a sensible and efficient sequence of executing the tests.

For example, it may be that several high-priority tests are dependent on a single low-priority test to set up essential data or starting conditions. In that case, the low-priority test should be executed before the high-priority tests, even when risk priority is the most important factor.

The same logic may apply where tests are dependent on other tests. Actually, this is one reason why tests should ideally be designed to be independent of any other tests. Independent tests can be run in any order.

In Agile development, rapid feedback from tests is key to efficient working of the team, but this means that the test execution schedule is biased toward confirmation tests and short tests, which can be run quickly. This also shows why it is important to be able to identify (or tag) tests so that they can be assembled into different test suites for different execution schedules, something that is particularly important for automated tests.

In Agile development, it is more likely to be the tester or the team rather than the test manager who is making the test execution schedule, but whoever does it, they need to balance the trade-off between efficiency, priority of the tests and the objective of the test execution at the time.

This learning objective is a K3 in the Syllabus, which means that you need to be able to produce a test execution schedule, taking priorities and dependencies into account. Refer to the exercise at the end of this chapter, and the mock exam in Chapter 7.

5.1.6 Test pyramid

The **test pyramid** is a metaphorical model, shown in Figure 5.1, which helps people think about how to focus test automation effort. It consists of a varying number of layers. The one shown in this figure is aligned with the test levels defined in the ISTQB syllabus. If you search online, you will also see one where there are only three layers, unit tests, service tests and UI tests.

Test pyramid A graphical model representing the relationship of the amount of testing per level, with more at the bottom than at the top.

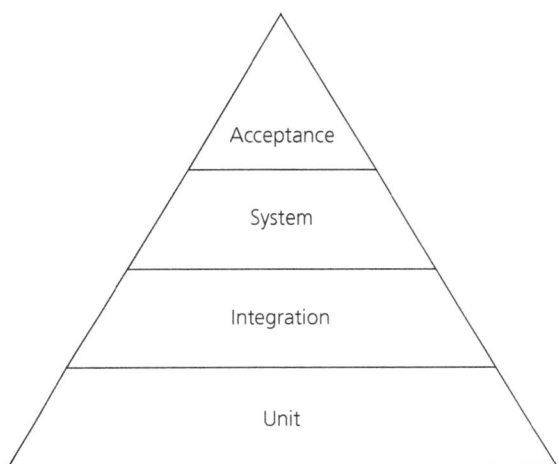

FIGURE 5.1   The test pyramid

Whichever version of this model you look at, the idea is that, at the lowest layer of the model (i.e. unit tests, aka component tests), the tests are fine-grained. That is, they are focused on a small piece of functionality, they are small, they run quickly and (if designed properly) can run independently, isolated from other parts of the system. All of these properties also tend to make such tests resilient and easier to maintain.

As you move up the layers, the tests become coarser grained. They cover larger pieces of functionality, they are larger tests, they run more slowly and they are dependent on more and more parts of the system. Therefore, these tests become more brittle and harder to maintain.

So why is it a pyramid rather than just a spectrum? Because the point was that we should have most of our testing, especially our automated testing, concentrated at the base of the pyramid—that's why it's wider. It's easier to automate through APIs, service layers, and the code itself, which is what we do with unit and integration level tests. System and acceptance level tests automate through a user interface, typically, and that's harder to do, much harder to do well, and often expensive to maintain, especially if care is not taken to design for maintainability.

The test pyramid is a metaphor; it's not to be taken too literally. If you are saying something like, 'We have 10,000 unit tests, 1,000 integration tests, 100 system tests, and 10 acceptance tests, therefore we're doing the pyramid perfectly,' you're thinking about this in the wrong way. To point out just one thing that's wrong with that quantitative thinking, consider the number of lines of code covered by a unit test compared to a system test. A unit test may cover 20 or 30 lines of code. A system test may cover 2,000, 3,000, or even more lines of code.

The test pyramid is really just a way of restating a basic principle of testing that we covered earlier: early QA and testing are helpful because they filter out defects earlier and cheaper. That way, when we get to the later parts of the project (in a sequential model) or iteration (in Agile), we don't have a lot of bugs left to discover and remove. Remember, if you're talking about a two-week iteration, you don't have time to be surprised by a lot of bugs found late.

5.1.7 Testing quadrants

Testing quadrants A classification model of test types/test levels in four quadrants, relating them to two dimensions of test objectives: supporting the product team versus critiquing the product, and technology-facing versus business-facing.

The **testing quadrants** concept was developed by Brian Marick as a way of thinking about the different types of testing that we do. It is used especially in an Agile context. An example, shown in Figure 5.2, puts the test levels and types defined in the Foundation syllabus into the context of the test quadrant.

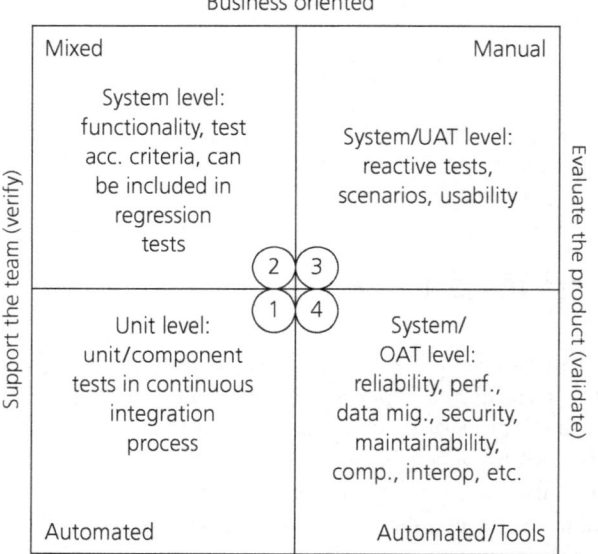

FIGURE 5.2   The test quadrants

The test quadrants are defined on two dimensions. The tests on the left side of the quadrant support the team. They are verification tests that check that the defined requirements are satisfied. The tests on the right side of the quadrant evaluate the product. They are validation tests that make sure that the product actually solves customer and user problems. The tests at the bottom are technology oriented. The tests at the top are business oriented.

Let's go around the quadrants, starting in quadrant one (bottom left). Quadrant one is where the automated unit tests, component integration tests and, in some cases, component tests would go (at least if component test means something different than unit test on your project). By the way, unit testing is the preferred term in Agile teams. Component test, even though it's defined as a synonym for unit test in the Foundation syllabus, is not used as a synonym for unit testing in most Agile teams that we've worked with. So, if we have separate tests for components, then those could be as part of quadrant one. Quadrant one tests are automated tests and are included in the continuous integration process.

Quadrant two (top left) are system level functional tests especially those focused on acceptance criteria, such as acceptance test-driven development tests. Quadrant two tests can be manual or automated. If they're automated, they can be included as part of the regression test pack, and possibly even incorporated within the continuous integration framework.

Quadrant three (top right) tests are manual tests at the system and user acceptance test level. This includes activities like exploratory testing and testing of usability.

Quadrant four (bottom left) tests are system tests or operational acceptance tests, including non-functional tests related to reliability, performance and security, maintainability, compatibility, interoperability across systems and data migration. Quadrant four tests are either automated or manual, but heavily supported by tools.

Each iteration can include tests that are in any of these quadrants.

Let's look at an example of how the test quadrants work for one of our clients, shown in Figure 5.3. First, we'll explain their lifecycle, and then we'll look at where the different quadrants fit in.

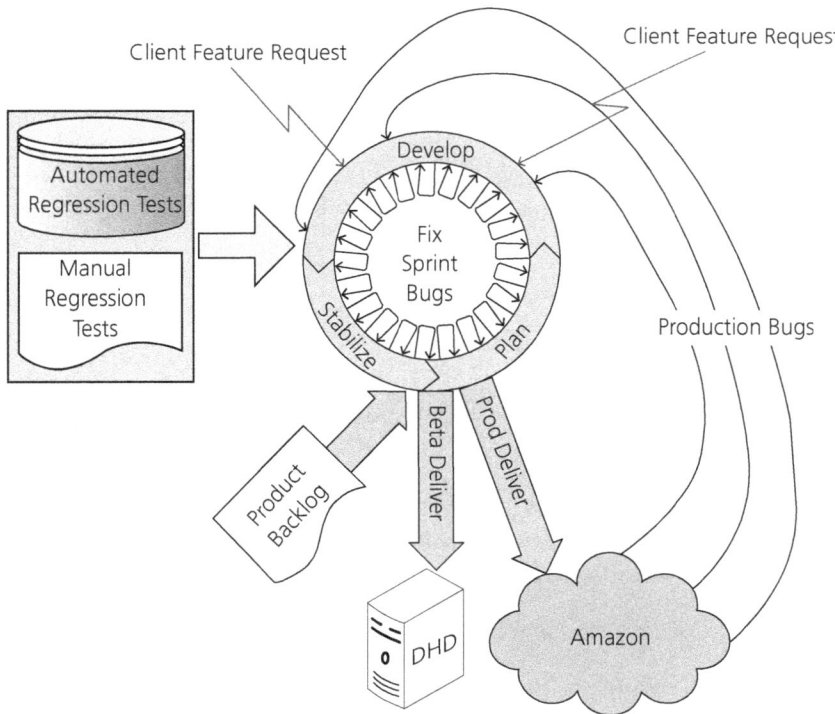

FIGURE 5.3   Test quadrants example

This organization follows a four-week long Scrum-based lifecycle. There's a planning period at the beginning of each iteration. There's a development period. Finally, there's a stabilization period at the end.

During iteration planning, the product backlog is examined, and the user stories for the sprint backlog are selected. They go into a development period, where the user stories are developed and tested.

Notice that, throughout the sprint, even during the planning period, sprint bugs are being fixed. In addition, during the development period, they allow something that theoretically isn't supposed to happen. During the sprint, the customers might well be finding bugs in the production system in the Amazon or DHD environments. The less urgent bugs are put into the product backlog, but the most urgent are inserted in the sprint backlog for the current iteration, even if reported during the stabilization period. This is forbidden in Scrum theory, but our client does it as a way of responding to customer needs. They do it in spite of the theory because their customers urgently need these bug fixes. Similarly, if a customer has urgent feature requests, those can become part of the iteration at any point prior to the stabilization period.

At the end of the iteration, after stabilization, they have a beta delivery to their DHD environment. This is used by a limited set of customers. If, after a few days, the beta release looks good, they deliver into their Amazon EC3 cloud environment. This production release happens during the next iteration.

How do the quadrants relate to what's happening? The quadrant one tests are the automated unit tests included in the continuous integration framework. These tests are developed and executed starting with development, of course, and continuing all the way through the stabilization period. The quadrant two tests are the feature verification tests, created as part of planning and during development, and executed during development and throughout stabilization. Quadrant three tests are used, in the form of exploratory testing, during development and stabilization.

Quadrant four tests, the non-functional types of tests, happen occasionally, but only when they're needed. There might be three or four sprints in a row where no Q4 non-functional tests occur. This is actually a mistake on their part, which one of us identified as part of a consulting engagement with them. We recommended that they adopt a better, risk-based approach to determining how frequently such non-functional tests were required.

5.2 RISK MANAGEMENT

> **SYLLABUS LEARNING OBJECTIVES FOR 5.2 RISK MANAGEMENT (K2)**
>
> FL-5.2.1   **Identify risk level by using risk likelihood and risk impact (K1)**
>
> FL-5.2.2   **Distinguish between project risks and product risks (K2)**
>
> FL-5.2.3   **Explain how product risk analysis may influence thoroughness and scope of testing (K2)**
>
> FL-5.2.4   **Explain what measures can be taken in response to analyzed product risks (K2)**

This section covers a topic that we believe is critical to testing: risk. Let's look closely at risks, the possible problems that might endanger the objectives of the project stakeholders, whether from internal or external sources. We will discuss how to determine the level of risk using likelihood and impact as part of risk analysis (including risk identification and risk assessment). We will see that there are risks related to the product and risks related to the project and look at typical risks in both categories. Finally, and most important, we'll look at various ways that we can use the results of risk analysis to exercise risk control (including risk mitigation and risk monitoring). Risk analysis and risk control allow us to carry out risk-based testing, where tests are selected, prioritized and managed using risk.

As you read this section, make sure to attend carefully to the Glossary terms **product risk**, **project risk**, **risk**, **risk analysis**, **risk assessment**, **risk control**, **risk identification**, **risk level**, **risk management**, **risk mitigation**, **risk monitoring** and **risk-based testing**.

5.2.1 Risk definition and risk attributes

Risk is a word we all use loosely, but what exactly is risk? Simply put, it is the possibility of a negative or undesirable outcome. In the future, a risk has some likelihood between 0% and 100%; it is a possibility, not a certainty. In the past, however, either the risk has materialized and become an outcome or issue, or it has not; the likelihood of a risk in the past is either 0% or 100%.

> **Risk** A factor that could result in future negative consequences.

The likelihood of a risk becoming an outcome is one factor to consider when thinking about the **risk level** associated with its possible negative consequences. The more likely the outcome is, the worse the risk. However, likelihood is not the only consideration.

> **Risk level** (risk exposure) The measure of a risk defined by risk impact and risk likelihood.

For example, most people are likely to catch a cold in the course of their lives, usually more than once. The typical healthy individual suffers no serious consequences. Therefore, the overall level of risk associated with colds is low for this person. But the risk of a cold for an elderly person with breathing difficulties would be high. The potential consequences or impact is an important consideration affecting the level of risk too.

Remember that in Chapter 1 we discussed how system context, and especially the risk associated with failures, influences testing. Here, we'll get into more detail about the concept of risks, how they influence testing and specific ways to manage risk.

5.2.2 Project risks and product risks

We can classify risks into project risks (factors relating to the way the work is carried out, including the test activities) and product risks (factors relating to what is produced by the work, that is, the thing we are testing). We will look at product risks first.

Product risks

You can think of a **product risk** as the possibility that the system or software might fail to satisfy some reasonable customer, user or stakeholder expectation. Unsatisfactory software might omit some key function that the customers specified, the users required or the stakeholders were promised. It might be unreliable and frequently fail to behave normally. It might fail in ways that cause financial or other damage to a user or the company that user works for. It might have problems with data integrity and data quality, such as data migration issues, data conversion problems, data transport problems and violations of data standards such as field and referential integrity. It might have problems related to a particular quality characteristic. The problems might not involve functionality, but rather security, reliability, usability, maintainability or performance. This type of quality attribute-related risk is sometimes referred to as a 'quality risk'. Generally, the software could fail to perform its intended functions, dissatisfying users and customers. If you can write a test for it, it's a product risk.

> **Product risk** A risk that impacts the quality of a product.

Here are some example product risks:

- Software might not perform its intended functions according to the specification.
- Software might not perform its intended functions according to user, customer and/or stakeholder needs or expectations (which is usually different from the first!).
- A particular computation may be performed incorrectly in some circumstances.
- A loop control structure may be coded incorrectly.
- Response times may be inadequate for a high-performance transaction processing system.
- User experience (UX) feedback might not meet product expectations.

If a product risk becomes an actual outcome (i.e. one or more bugs in the software that escape to production and result in failure[s] in the software), the consequences can range from non-existent to minor to major to catastrophic. At the very least, such failures, when noticed, result in load on the support staff and incur higher costs to repair than had they been fixed before release. Going up the consequences scale, unhappy users may damage the organization's reputation through online reviews or chat channels. The organization could lose money. Other third-party organizations could be affected, which could give them a cause of action to sue. In severe cases, the escape of such a defect, especially if associated with negligence by the development team, could result in someone going to jail. Of course, the worst case is when a safety-critical system or piece of software fails, damaging objects, injuring people, or even killing someone.

Project risks

We just discussed risks to product quality. However, testing is an activity like the rest of the project and thus it is subject to risks that endanger the project. To deal with the **project risks** that apply to testing, we can use the same concepts we apply to identifying, prioritizing and managing product risks.

Project risk A risk that impacts project success.

Remembering that a risk is the possibility of a negative outcome, what project risks affect testing? There are direct risks such as the late delivery of the test items to the test team or availability issues with the test environment. There are also indirect risks such as excessive delays in repairing defects found in testing or problems with getting professional system administration support for the test environment.

Of course, these are merely four examples of project risks; many others can apply to your testing effort. To discover these risks, ask yourself and other project participants and stakeholders: 'What could go wrong on the project to delay or invalidate the test plan, the test strategy and the test estimate? What are unacceptable outcomes of testing or in testing? What are the likelihoods and impacts of each of these risks?' This process is very much like the risk analysis process for products. Checklists and examples can help you identify test project risks [Black 2004].

Here are some project risks, in various categories, along with some examples and ways to deal with them.

Organizational issues:

- Delays may occur in delivery, task completion or satisfaction of exit criteria or definition of done. For example, logistics or product quality problems may block tests. These can be mitigated through careful planning, good defect triage and management and robust test design.

- Inaccurate estimates, reallocation of funds to higher priority projects, or general cost cutting across the organization may result in inadequate funding.
- Late changes may result in substantial rework. For example, excessive change to the product may invalidate test results or require updates to test cases, expected results and environments. These can be mitigated through good change control processes, robust test design and lightweight test documentation. When severe defects occur, transference of the risk by escalation to management is often in order.

People issues:

- Skills, training and staff may not be sufficient. For example, shortages of people, skills or training may lead to problems with communicating and responding to test results, unrealistic expectations of what testing can achieve and needlessly complex project or team organization.
- Personnel issues may cause conflict and problems, affecting work. For example, if two people do not get along, working against each other rather than towards a common goal (it happens), it is demoralizing for the whole team as well as degrading the quality of everyone's work.
- Users, business staff or subject matter experts may not be available due to conflicting business priorities.
- Testers may not communicate their needs and/or the test results adequately. For example, high-level managers may not realize the need for ongoing support, time and effort for test automation after an initial start, if testers and automaters do not communicate the need for this.
- Developers and/or testers may fail to follow up on information found in testing and reviews (for example, not improving development and testing practices).
- There may be an improper attitude toward, or expectations of, testing (for example, not appreciating the value of finding defects during testing).

Technical issues:

- Requirements or user stories may not be clear enough or well enough defined. For example, ambiguous, conflicting or unprioritized requirements, an excessively large number of requirements given other project constraints, high system complexity and quality problems with the design, the code or the tests mean that the system will take longer to develop and may not meet customer expectations. The best cure for this is clear unambiguous requirements.
- The requirements may not be met, given existing constraints. For example, a requirement may want the app to be able to check the user's bank balance to see if they can pay for what they are ordering, but the bank software does not allow it.
- The test environment may not be ready on time or may not be adequate. For example, insufficient or unrealistic test environments may yield misleading results, including false positives and false negatives. One option is to transfer the risks to management by explaining the limits on test results obtained in limited environments. Mitigation, sometimes complete alleviation, can be achieved by outsourcing tests such as performance tests that are particularly sensitive to proper test environments, or by using virtualization.

- Data conversion, migration planning and their tool support may be late.
- Weaknesses in the development process may impact the consistency or quality of project work products such as design, code, configurations, test data and test cases. For example, test items may not install in the test environment. This can be mitigated through smoke (or acceptance) testing prior to starting other testing or as part of a nightly build or continuous integration. Having a defined uninstall process is a good contingency plan.
- Poor defect management and similar problems may result in accumulated defects and other technical debt.

Supplier issues:

- A third party may fail to deliver a necessary product or service or go bankrupt.
- Contractual issues may cause problems to the project. For example, problems with underlying platforms or hardware, failure to consider testing issues in the contract, or failure to properly respond to the issues when they arise can quickly add up to serious delays as well as time-consuming negotiations with a supplier. The best way to mitigate this is to ensure that the contract is solid to begin with, and have the technical details reviewed by testers or a test manager.

Project risks do not just affect testing of course, they also affect development. In some organizations, project managers are responsible for dealing with all project risks, but sometimes testers or test managers are given responsibility for test-related project risks (as well as product risks).

There may be other risks that apply to your project and not all projects are subject to the same risks. Refer to Chapter 2 of Black [2009] and Chapters 6 and 7 of Black [2004] for a discussion on managing project risks during testing and in the test plan.

Finally, don't forget that test items can also have risks associated with them. For example, there is a risk that the test plan will omit tests for a functional area or that the test cases do not exercise the critical areas of the system.

5.2.3 Product risk analysis

Risk analysis

Risk-based testing starts with product **risk analysis**, which includes both **risk identification** and **risk assessment**. One technique for product risk analysis is a close reading of the requirements specification, user stories, design specifications, user documentation and other items. Another technique is brainstorming with many of the project stakeholders. Another is a sequence of one-to-one or small group sessions with the business and technology experts in the company. Some people use all these techniques when they can. To us, a team-based approach that involves the key stakeholders and experts is preferable to a purely document-based approach, as team approaches draw on the knowledge, wisdom and insight of the entire team to determine what to test and how much.

While you could perform the risk analysis by asking 'What should we worry about?', usually more structure is required to avoid missing things. One way to provide that structure is to look for specific risks in particular product risk categories. You could consider risks in the areas of functionality, localization, usability, reliability, performance and supportability. Alternatively, you could use the quality characteristics and sub-characteristics from ISO/IEC 25010 [2011], as

Risk analysis The overall process of risk identification and risk assessment.

Risk-based testing Testing in which the management, selection, prioritization and use of testing activities and resources are based on corresponding risk types and risk levels.

Risk identification The process of finding, recognizing and describing risks.

Risk assessment The process to examine identified risks and determine the risk level.

each sub-characteristic that matters is subject to risks that the system might have troubles in that area. You might have a checklist of typical or past risks that should be considered. You might also want to review the tests that failed and the bugs that you found in a previous release or a similar product. These lists and reflections serve to jog the memory, forcing you to think about risks of particular types, as well as helping you structure the documentation of the product risks.

When we talk about specific risks, we mean a particular kind of defect or failure that might occur. For example, if you were testing the calculator utility that is bundled with Microsoft Windows, you might identify incorrect calculation as a specific risk within the category of functionality. However, this is too broad. Consider incorrect addition. This is a high-impact kind of defect, as everyone who uses the calculator will see it. It is unlikely, since addition is not a complex algorithm. Contrast that with an incorrect sine calculation. This is a low-impact kind of defect since few people use the sine function on the Windows calculator. It is more likely to have a defect, though, since sine functions are hard to calculate.

It's worth making the point here that in early risk analysis, we are making educated guesses. Some of those guesses will be wrong. Make sure that you plan to re-assess and adjust your risks at regular intervals in the project and make appropriate course corrections to the testing or the project itself.

One common problem people have when organizations first adopt risk-based testing is a tendency to be excessively alarmed by some of the risks once they are clearly articulated. Don't confuse impact with likelihood or vice versa. You should manage risks appropriately, based on likelihood and impact. Triage the risks by understanding how much of your overall effort can be spent dealing with them.

It's very important to maintain a sense of perspective, a focus on the point of the exercise. As with life, the goal of risk-based testing should not be, and cannot practically be, a risk-free project. What we can accomplish with risk-based testing is the marriage of testing with best practices in risk management to achieve a project outcome that balances risks with quality, features, budget and schedule.

Assigning a risk level

After identifying the risk items, you and, if applicable, the stakeholders, should review the list to assign the likelihood of problems and the impact of problems associated with each one. There are many ways to go about this assignment of likelihood and impact. You can do this with all the stakeholders at once. You can have the business people determine impact and the technical people determine likelihood, and then merge the determinations. Either way, the reason for identifying risks first and then assessing their level, is that the risks are relative to each other.

The scales used to rate likelihood and impact vary. Some people rate them high, medium and low. Some use a 1–10 scale. The problem with a 1–10 scale is that it is often difficult to tell a 2 from a 3 or a 7 from an 8 unless the differences between each rating are clearly defined. A five-point scale (very high, high, medium, low and very low) tends to work well.

Given two classifications of risk levels, likelihood and impact, we have a problem. We need a single, aggregate risk rating to guide our testing effort. As with rating scales, practices vary. One approach is to convert each risk classification into a number and then either multiply or add the numbers to calculate a risk priority number. For example, suppose a particular risk has a high likelihood and a medium impact. The risk priority number would then be six (2×3). This is described as a quantitative approach in the syllabus. You can also use what the syllabus describes as a qualitative approach, which is where a lookup table is used to assign an aggregate risk level. We have seen both approaches used successfully.

5.2.4 Product risk control

Dealing with risks within an organization is known as risk management, and testing is one way of managing aspects of risk. For any risk, product or project, you have four typical options:

- **Mitigate:** Take steps in advance to reduce the likelihood (and possibly the impact) of the risk.
- **Contingency:** Have a plan in place to reduce the impact should the risk become an outcome.
- **Transfer:** Convince some other member of the team or project stakeholder to reduce the likelihood or accept the impact of the risk.
- **Accept:** Do nothing about the risk, which is usually a smart option only when there's little that can be done or when the likelihood and impact are low.

There is another typical risk management option, buying insurance, which is not usually pursued for project or product risks on software projects, though it is not unheard of.

Risk management activities include:

Risk management
The process for
handling risks.

- Identifying and analyzing (and re-evaluating on a regular basis) what can go wrong.
- Determining the priority of risks, which risks are the most important to deal with.
- Implementing actions to mitigate those risks, to reduce their likelihood or impact or both.
- Making contingency plans to deal with risks if they do happen.

We can deal with test-related risks to the project and product by applying some straightforward, structured risk management techniques. The first step is to assess or analyze risks early in the project. Like a big ocean liner, projects, especially large projects, require steering well before the iceberg is in plain sight. By using a test plan, you can remind yourself to consider and manage risks during the planning activity.

Controlling product risks through risk-based testing

Risk-based testing is the idea that we can organize our testing efforts in a way that reduces the residual level of product risk when the system is delivered. Risk-based testing uses risk to prioritize and emphasize the appropriate tests during test execution, but it is about more than that. Risk-based testing starts early in the project, identifying risks to system quality and using that knowledge of risk to guide testing planning, specification, preparation and execution. In addition to the risk identification and risk assessment (i.e. risk analysis) we discussed earlier, risk-based testing also involves both **risk mitigation** (testing to provide opportunities to reduce the likelihood of defects, especially high-impact defects) and contingency (testing to identify workarounds to make the defects that do get past us less painful).

Risk mitigation The
process through which
decisions are reached
and protective measures
are implemented for
reducing or maintaining
risks to specified levels.

Risk-based testing also involves **risk monitoring**, which is determining how well we are doing at finding and removing defects in critical areas, as is shown later in Table 5.2. Risk-based testing can also involve using risk analysis to identify proactive opportunities to remove or prevent defects through non-testing activities and to help us select which test activities to perform. Collectively, the syllabus refers to product risk mitigation and product risk monitoring as **risk control**.

Mature test organizations use testing to reduce the risk associated with delivering the software to an acceptable level [Beizer 1990, Hetzel 1988]. In the middle of the 1990s, a number of testers, including us, started to explore various techniques for risk-based testing. In doing so, we adapted well-accepted risk management concepts to software testing. Applying and refining risk assessment and management techniques are discussed in Black [2004] and Black [2009]. For an alternative view, refer to Chapter 11 of Pol *et al.* [2002]. The origin of the risk-based testing concept can be found in Chapter 1 of Beizer [1990] and Chapter 2 of Hetzel [1988].

Risk monitoring The activity that checks and reports the status of known risks to stakeholders.

Risk control The overall process of risk mitigation and risk monitoring.

Mitigation options

Armed with a risk priority number, we can now decide on the various risk mitigation options available to us. Do we use formal training for developers or analysts, rely on cross-training and reviews or assume they know enough? Do we perform extensive testing, cursory testing or no testing at all? Should we ensure unit testing and system testing coverage of this risk? These options and more are available to us.

As you go through this process, make sure you capture the key information in a document. We are not fond of excessive documentation, but this quantity of information simply cannot be managed in your head. We recommend a lightweight table like the one shown in Table 5.1, which we usually capture in a spreadsheet (although a few modern test management tools now provide native support for risk-based testing or can be customized to support it).

TABLE 5.1 A risk analysis template

Product risk	Likelihood	Impact	Risk priority number	Mitigation
Risk category 1				
Risk 1				
Risk 2				
Risk n				

Concluding thoughts on risk

To summarize, the results of product risk analysis are used for the following purposes:

- To determine the scope of testing, both at a macro level (i.e. the overall breadth of coverage across types and levels) and at a micro level (i.e. the depth of testing within the relevant types and levels).

- To determine the test techniques to be used and the coverage needed.

- To determine the particular levels and types of testing to be performed (for example, functional testing, security testing, performance testing).

- To estimate the test effort, at a task level, based on the extent of testing to be carried out for the different levels and types of testing.

- To prioritize testing in order to find the most critical defects as early as possible.

- To determine whether any activities in addition to testing could be employed to reduce risk, such as providing training in design and testing to inexperienced developers.

Let's finish this section with two quick tips about product risk analysis. First, remember to consider both likelihood and impact. While it might make you feel like a hero to find lots of defects, testing is also about building confidence in key functions. We need to test the things that probably will not break but would be catastrophic if they did.

Second, we reiterate that in early risk analysis, we are making educated guesses. Make sure that you follow up and revisit the risk analysis at key project milestones. For example, if you are following a sequential model such as the V-model, you might perform the initial analysis during the requirements phase, then review and revise it at the end of the design and implementation phases, as well as prior to starting unit test, integration test and system test. We also recommend revisiting the risk analysis during testing. You might find you have discovered new risks or found that some risks were not as risky as you thought.

5.3 TEST MONITORING, TEST CONTROL AND TEST COMPLETION

> **SYLLABUS LEARNING OBJECTIVES FOR 5.3 TEST MONITORING, TEST CONTROL AND TEST COMPLETION (K2)**
>
> **FL-5.3.1**   **Recall metrics used for testing (K1)**
>
> **FL-5.3.2**   **Summarize the purposes, contents and audiences for test reports (K2)**
>
> **FL-5.3.3**   **Exemplify how to communicate the status of testing (K2)**

In this section, we'll review techniques and metrics that are commonly used for monitoring test implementation and execution. We'll focus especially on the use and interpretation of such test metrics for reporting, controlling and analyzing the test effort, including those based on defects and those based on test data. We'll also look at options for reporting test status using such metrics and other information.

As you read, remember to watch for the Glossary terms **test control**, **test monitoring**, **test progress report** and **test completion report**.

Test monitoring is concerned with gathering data and information about test activities. We use these data and information to evaluate progress on testing and against the exit criteria, especially in terms of coverage, defects and test status. Test monitoring involves deciding which metrics should be gathered and how that information should be communicated in test reports.

Test control is using the insights gained from test monitoring to guide or control the remaining testing in the most effective and efficient fashion. Test control can take the form of reprioritizing tests based on changes in risk status, re-evaluating the status of a test item with respect to entry or exit criteria after changes are made, changing the test schedule (e.g. due to delays in resources or delivery of test items), adding new resources, or other actions.

Projects do not always unfold as planned. In fact, any human endeavour more complicated than a family picnic is likely to vary from plan. Risks become occurrences. Stakeholder needs evolve. The world around us changes. When plans and reality diverge, we must act to bring the project back under control, and testing is no exception.

In some cases, the test findings themselves are behind the divergence; for example, suppose the quality of the test items proves unacceptably bad and delays test progress. In other cases, testing is affected by outside events; for example, testing can be delayed when the test items show up late or the test environment is unavailable. Test control is about guiding and corrective actions to try to achieve the best possible outcome for the project.

The specific corrective or guiding actions depend, of course, on what we are trying to control. Consider the following hypothetical examples of test control actions:

- A portion of the software under test will be delivered late, after the planned test start date. Market conditions dictate that we cannot change the release date. Test control might involve re-prioritizing the tests so that we start testing against what is available now.

- For cost reasons, performance testing is normally run on weekday evenings during off-hours in the production environment. Due to unanticipated high demand for your products, the company has temporarily adopted an evening shift that keeps the production environment in use 18 hours a day, five days a week. Test control might involve rescheduling the performance tests for the weekend.

- The item being tested had previously met its entry (or exit) criteria, but since the criteria were evaluated, the test item has changed due to defects found and resulting rework. Control actions may include re-evaluating the entry (or exit) criteria.

While these examples show test control actions that affect testing, the project team might also have to take some actions that affect others on the project. For example, suppose that the test completion date is at risk due to a high number of defect fixes that fail confirmation testing in the test environment. In this case, test control might involve requiring the developers making the fixes to thoroughly retest the fixes prior to checking them in to the code repository for inclusion in a test build.

As test activities are completed, e.g. when a test level is finished, when an iteration or sprint concludes, when software or a system is released, etc.—test completion activities occur as a final step. These are important to capture experience or knowledge

Test monitoring The activity that checks the status of testing activities, identifies any variances from planned or expected, and reports status to stakeholders.

Test control The activity that develops and applies corrective actions to get a test project on track when it deviates from what was planned.

gained, to archive testware for reuse, to capture data and metrics that can be used for future estimation efforts and to otherwise capitalize on the work done for the future.

5.3.1 Metrics used in testing

The purpose of metrics in testing

Having developed our plans, defined our test strategies and approaches and estimated the work to be done, we must now track our testing work as we carry it out. Test monitoring can serve various purposes during the project, including the following:

● Give the test team and the test manager feedback on how the testing work is going, allowing opportunities to guide and improve the testing and the project or the iteration. This would be done by collecting time and cost data about progress versus the planned schedule and budget.

● Provide the project team with visibility about the test results and the quality of the test object.

● Measure the status of the testing, test coverage and test items against the exit criteria to determine whether the test work is done, and to assess the effectiveness of the test activities with respect to the objectives.

● Gather data for use in estimating future test efforts, including the adequacy of the test approach.

● Providing any other data and information needed to support test control and completion.

Especially for small projects, the test manager or a delegated person can gather test progress monitoring information manually using documents, spreadsheets and simple databases. When working with large teams, distributed projects and long-term test efforts, we find that the efficiency and consistency of data collection is aided by the use of automated tools (refer to Chapter 6).

Common test metrics

We will show some examples of using metrics in testing and conclude this section with a list of other common test metrics.

In Figure 5.4 columns A and B show the test ID and the test case or test suite name. The state of the test case is shown in column C ('Warn' indicates a test that resulted in a minor failure). Column D shows the tested configuration, where the codes A, B and C correspond to test environments described in detail in the test plan. Columns E and F show the defect (or bug) ID number (from the defect tracking database) and the risk priority number of the defect (ranging from 1, the highest risk, to 25, the least risky). Column G shows the initials of the tester who ran the test. Columns H to L capture data for each test related to dates, effort and duration (in hours). We have metrics for planned and actual effort and dates completed which would allow us to summarize progress against the planned schedule and budget. This spreadsheet can also be summarized in terms of the percentage of tests which have been run and the percentage of tests which have passed and failed. Column M is for comments ('out of runway' means that the tests for this aspect will no longer be run because we are out of time).

Figure 5.4 might show a snapshot of test progress during the test execution period, or perhaps even at test closure if it were deemed acceptable to skip some of the tests.

During the analysis, design and implementation of the tests, such a worksheet would show the state of the tests in terms of their state of development.

	Test ID	Test Suite/Case	Status	System Config	Bug ID	Bug RPN	Run By	Plan Date	Act Date	Plan Effort	Actual Effort	Test Duration	Comment
1			System Test Case Summary										
2			Cycle One										
3													
7	1.000	*Functionality*											
8	1.001	File	Fail	A	701	1	LTW	1/8	1/8	4	6	6	
9	1.002	Edit	Fail	A	709	1	LTW	1/9	1/10	4	8	8	
10					710	5							
11					718	3							
12					722	4							
13	1.003	Font	Pass	B			IHB	1/10	1/10	4	4	4	
14	1.004	Tables	Warn	B	708	15	IHB	1/8	1/9	4	5	5	
15	1.005	Printing	Skip					1/10		4			Out of runway
16		**Suite Summary**						**1/10**	**1/10**	**20**	**23**	**23**	
17													
18	2.000	*Performance/Stress*											
19	2.001	Solaris Server	Warn	A,B,C	701	1	EM	1/10	1/13	4	8	24	Replan 1/11
20	2.002	Windows Server	Fail	A,B,C	724	2	EM	1/11	1/14	4	4	24	Replan 1/12
21					713	2							
22					725	1							
23	2.003	Linux Server	Skip					1/12		4			Out of runway
24		**Suite Summary**						**1/12**	**1/14**	**12**	**12**	**48**	
25													
26	3.000	*Error Handling/Recovery*											
27	3.001	Corrupt File	Fail	A	701	1	LTW	1/8	1/9	4	8	8	
28					706	2							
29					707	4							
30					709	1							
31					710	5							
32					713	2							
33	3.002	Server Crash	Fail	A	712	6	LTW	1/9	1/10	4	6	6	
34					713	2							
35					717	1							
36		**Suite Summary**						**1/9**	**1/10**	**8**	**14**	**14**	
37													
38	4.000	*Localization*											
39	4.001	Spanish	Skip										
40	4.002	French	Skip										
41	4.003	Japanese	Skip										
42	4.004	Chinese	Skip										
43		**Suite Summary**						**1/0**	**1/0**	**0**	**0**	**0**	

FIGURE 5.4 Test case summary worksheet

In addition to test case status, it is also common to monitor test progress during the test execution period by looking at the number of defects found and fixed. Figure 5.5 shows a graph that plots the total number of defects opened and closed over the course of the test execution so far.

It also shows the planned test period end date and the planned number of defects that will be found. Ideally, as the project approaches the planned end date, the total number of defects opened will settle in at the predicted number and the total number of defects closed will converge with the total number opened. These two outcomes tell us that we have found enough defects to feel comfortable that we have finished, that we have no reason to think many more defects are lurking in the product and that all known defects have been resolved.

Charts such as Figure 5.5 can also be used to show failure rates or defect density. When reliability is a key concern, we might be more concerned with the frequency with which failures are observed than with how many defects are causing the failures.

In organizations that are looking to produce ultra reliable software, they may plot the number of unresolved defects normalized by the size of the product, either in thousands of source lines of code (KSLOC), function points (FP) or some other metric of code size. Once the number of unresolved defects falls below some predefined threshold—for example, three per million lines of code—then the product may be deemed to have met the defect density exit criteria.

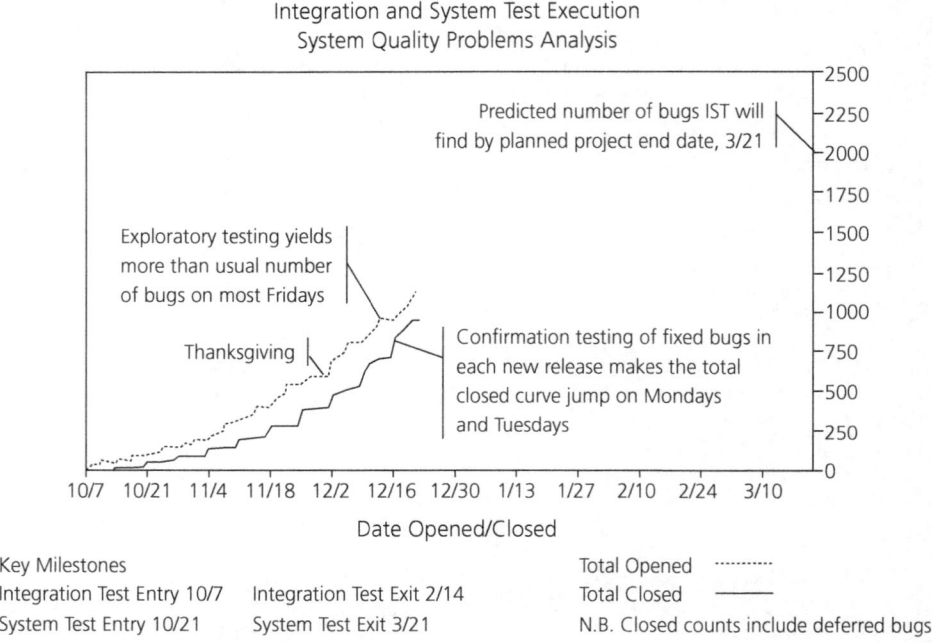

FIGURE 5.5 Total defects opened and closed chart

Measuring test progress based on defects found and fixed is common and useful, if used with care. Avoid using defect metrics alone, as it is possible to achieve a flat defect find rate and to fix all the known defects by stopping any further testing, by deliberately impeding the reporting of defects and by allowing developers to reject, cancel or close defect reports without any independent review.

That said, test progress monitoring techniques vary considerably depending on the preferences of the testers and stakeholders, the needs and goals of the project, regulatory requirements, time and money constraints and other factors.

Common test metrics monitoring include:

● Metrics that measure the progress of the test effort as a part of the project, such as the number of tasks completed, the effort expended, etc.

● Metrics that measure the progress of the test process itself, such as the percentage of planned test cases or test environments which are ready, the status of the tests in terms of being ready for execution, passed, failed, or blocked, or the amount of time spent executing tests.

● Metrics that measure functional or non-functional product quality, such as the percentage of common use cases or activities which work properly (functional) or the mean time between failure (non-functional).

- Metrics that measure defect status, such as trend graphs showing defect find/fix totals over time, a rolling average of the team's defect detection percentage, defect density, confirmation test results, etc.

- Metrics that measure risk, such as the number of risks identified, not tested, partially tested, fully tested but with open defects and tested without open defects (i.e. measures of the residual risk to the quality of the product).

- Metrics of the relevant test basis coverage, such as requirements or risk coverage for system level tests or code coverage for unit level tests.

- The economics of testing, such as the costs and benefits of continuing test execution in terms of finding the next defect or running the next test, the cost of quality, or the total test cost.

As a complementary monitoring technique, you might assess the subjective level of confidence the testers have in the test items. However, avoid making important decisions based on subjective assessments alone, as people's impressions have a way of being inaccurate and coloured by bias.

5.3.2 Purpose, contents and audiences for test reports

The purpose of test reports

Test monitoring is about gathering detailed test data; reporting test progress is about effectively communicating our findings to other project stakeholders. As with test progress monitoring, in practice there is wide variability observed in how people report test progress, with the variations driven by the preferences of the testers and stakeholders, the needs and goals of the project, regulatory requirements, time and money constraints and limitations of the tools available for test progress reporting. Often variations or summaries of the metrics used for test progress monitoring, such as Figure 5.4 and Figure 5.5 are used for test progress reporting, too. Regardless of the specific metrics, charts and reports used, test progress reporting is about helping project stakeholders understand the results of a test period, especially as it relates to key project goals and whether (or when) exit criteria were satisfied.

In addition to notifying project stakeholders about test results, test progress reporting is often about enlightening and influencing them. This involves analyzing the information and metrics available to support conclusions, recommendations and decisions about how to guide the project forward or to take other actions. For example, we might estimate the number of defects remaining to be discovered, present the costs and benefits of delaying a release date to allow for further testing, assess the remaining product and project risks and offer an opinion on the confidence the stakeholders should have in the quality of the system under test.

You should think about test progress reporting during test planning, since you will often need to collect specific metrics during and at the end of a test period to generate the test progress reports in an effective and efficient fashion. The specific data you will want to gather will depend on your specific reports, but common considerations include the following:

- How will you assess the adequacy of the test objectives for a given test level and whether those objectives were achieved?

- How will you assess the adequacy of the test approaches taken and whether they support the achievement of the project's testing goals?

- How will you assess the effectiveness of the testing with respect to these objectives and approaches?

For example, if you are doing risk-based testing, one main test objective is to subject the important product risks to the appropriate extent of testing. Table 5.2 shows an example of a chart that would allow you to report your test coverage and unresolved defects against the main product risk areas you identified in your risk analysis. If you are doing requirements-based testing, you could measure coverage in terms of requirements or functional areas instead of risks.

TABLE 5.2   Risk coverage by defects and tests

Product risk areas	Unresolved defects		Test cases to be run		
	Number	%	Planned	Actual	%
Performance, load, reliability	304	28	3,843	1,512	39
Robustness, operations, security	234	21	1,032	432	42
Functionality, data, dates	224	20	4,744	2,043	43
Use cases, user interfaces, localization	160	15	498	318	64
Interfaces	93	8	193	153	79
Compatibility	71	6	1,787	939	53
Other	21	2	760	306	40
	1,107	100	12,857	5,703	44

The content of test reports

Test progress report (test status report) A type of periodic test report that includes the progress of test activities against a baseline, risks, and alternatives requiring a decision.

Test completion report (test summary report) A type of test report produced at completion milestones that provides an evaluation of the corresponding test items against exit criteria.

The Syllabus discusses two types of test report: a **test progress report** and a **test completion report**. Both reports contain a lot of information in common; the main difference is that test progress reports are used at regular intervals throughout the project, during test activities, and would result in test control actions; the test completion report is in a sense the final test progress report for the whole project or iteration and is prepared at the end of a test activity, test level, or iteration. For example, a test completion report could also be used in a retrospective.

Because test progress reports are reporting on more dynamic information, there are some things which are included in this report that are no longer relevant for a test completion report, including:

● The current status of the test activities and progress against the test plan.
● Factors that are currently impeding the progress of the testing.
● The testing planned for the next period, to be included in the next test progress report.
● The quality of the test object as currently assessed by the testing.

When exit criteria are met, a tester or the test manager issues the test completion report. Such a report, created either at a key milestone or at the end of a test level, describes the results of a given test level or an iteration. In addition to including the kind of charts and tables shown earlier (but now at the end of the time period being monitored), you might discuss important events (especially problematic ones) that occurred during testing, the objectives of testing and whether they were achieved, the test approach, strategies followed, how well they worked and the overall effectiveness of the test effort. The test completion report draws on the test progress reports over the duration of the project or test level, particularly the last one.

The contents of both test progress reports and test completion reports may include the following:

- A summary of the testing performed.
- Information about what occurred during the period the report covers.
- Deviations from plan, with regard to the schedule, duration or effort of test activities.
- The status of the testing and of product (test object) quality with respect to the exit criteria or definition of done.
- The factors that have blocked or continue to block progress.
- Relevant metrics, such as those relating to defects, test cases, test coverage, test activity progress and resource consumption, as described in Section 5.3.1.
- New, changed and/or residual risks (refer to Section 5.2).
- Reusable test work products produced.

Not every test report will contain all this information (for example, a quick software update), and some may include additional information (for example, for a complex project with many stakeholders or under legal and regulatory requirements). The content of the report depends on the project, organizational requirements and the software development life cycle. For example, in Agile development, test progress reporting may be incorporated into task boards, defect summaries and burndown charts, which may be discussed during a daily stand-up meeting.

The standard ISO/IEC/IEEE 29119-3 [2021] gives structures and examples of test progress reports (referred to there as test status reports) and test completion reports.

5.3.3 Communicating the status of testing

In addition to the factors already mentioned that influence the content of a test report (either progress or completion), each test report should be tailored for the intended audience of the report. For example, a test completion report which is intended for the CEO and other high-level management should be short and concise, with overview summaries of the main points. It may contain a summary of defects of high and other priority levels and the percentage of tests passed, but information about budget and schedule would be of more importance to them. A test progress report which is intended for testers and developers would include detailed information about defect types and trends, such as we saw in Figure 5.5

How this information is best communicated to the audience varies significantly. Regulatory or compliance standards, which exist in industries as different as medical devices, defence industry and video gaming, may determine how some information is gathered and presented (not to mention what information must be gathered). Agile teams, as mentioned earlier, tend to have daily stand up meetings where such information would be reported, as well as retrospectives where test completion reports and activities would be highly relevant. The organizational test strategy may specify how reporting is to be done.

With those factors in mind, options for communicating this information include:

- Verbally, with the team or to other stakeholders (strongly favoured in Agile teams).
- Dashboards produced by various project management, test management and/or configuration management tools.
- Electronic communication tools, such as email or instant messaging.
- Documentation posted online, often generated by a project management, test management and/or configuration management tool.
- Formal test reports such as a test progress report or test completion report.

These options are not mutually exclusive. You can use multiple options, and probably should, depending on the audience. Senior managers may want formal reports (albeit concise ones), people working on the team in the same office may be fine with verbal communication, and distributed teams working offsite in different time zones may need to send and receive detailed written documents.

Remember to keep people's information needs and preferences front-of-mind when deciding how to deliver test information. As test professionals, we are often bearers of bad news, e.g. regression tests need to be re-run due to large-scale changes, a larger-than-expected number of bugs has been found, performance problems are due to a significant, hard-to-change database design decision made early in the project, etc. You will be fighting against confirmation bias when you tell people such bad news. Don't make your communication job even harder by just taking whatever pops out of your project management and/or test management tools and dropping those reports in front of every stakeholder, regardless of role or information needs.

5.4 CONFIGURATION MANAGEMENT

> **SYLLABUS LEARNING OBJECTIVE FOR 5.4 CONFIGURATION MANAGEMENT (K2)**
>
> **FL-5.4.1 Summarize how configuration management supports testing (K2)**

Configuration management A discipline applying technical and administrative direction and surveillance to identify and document the functional and physical characteristics of a configuration item, control changes to those characteristics, record and report change processing and implementation status, and verify that it complies with specified requirements.

In this brief section, we'll look at how configuration management relates to and supports testing. There are no Glossary terms specifically called out (and therefore examinable), but the Glossary does include the relevant term, **configuration management**.

Configuration management is a topic that often perplexes new practitioners, but, if you ever have the bad luck to work as a tester on a project where this critical activity is handled poorly, you will never forget how important it is. Briefly put, configuration management is in part about determining clearly what the items are that make up the software or system. These items include source code, test scripts, third-party software (including tools that support testing), hardware, data and both development and test documentation. Configuration management is also about making sure that these items are managed carefully, thoroughly and attentively throughout the entire project and product life cycle.

5.4.1 How configuration management supports testing

Configuration management has a number of important implications for testing. For one thing, it allows the testers to manage their testware and test results using the same configuration management mechanisms, as if they were as valuable as the source code and documentation for the system itself—which of course they are.

For another thing, configuration management supports the build process, which is essential for delivery of a test release into the test environment. It is critical to have a solid, reliable way of delivering test items that work and are the proper version.

Last but not least, configuration management allows us to map what is being tested to the underlying files and components that make it up. This is absolutely critical. For example, when we report defects, we need to report them *against* something, something which is configuration controlled or version controlled. If it is not clear what we found the defect in, the developers will have a very tough time of finding the defect in order to fix it. For the kind of test reports discussed earlier to have any meaning, we must be able to trace the test results back to what exactly we tested.

Configuration management for testing may involve ensuring the following:

- All test items of the test object are uniquely identified, version controlled, tracked for changes and related to each other, that is, what is being tested.
- All items of testware are uniquely identified, version controlled, tracked for changes, related to each other and related to a version of the test item(s) so that traceability can be maintained throughout the test process.
- All identified work products and software items are referenced unambiguously in test documentation.

Ideally, when testers receive an organized, version-controlled test release from a change-managed source code repository, it is accompanied by release notes which contain all the information shown.

In the last decade, as Agile became more mainstream and DevOps approaches became stable, a dazzling plethora of tools has emerged to support integration of configuration management, automated tests (at all levels), and the build process. These are often referred to as build pipelines or DevOps pipelines. These allow continuous integration of code and its associated automated tests as they're developed, with those tests (and perhaps more tests) executed to ensure a stable check-in. These can also support continuous delivery, where software with the latest updates (and the results of testing it) is available for testing shortly after the last check-in, and even continuous deployment, where the software is deployed (if the automated tests pass) to development, test and even (in some cases) production environments.

While our description was brief, configuration management is a topic that is as complex as test environment management. Advanced planning is critical to making this work. During the project planning stage, and perhaps as part of your own test plan, make sure that configuration management procedures and tools are selected. As the project proceeds, the configuration process and mechanisms must be implemented, and the key interfaces to the rest of the development process should be documented. Come test execution time, this will allow you and the rest of the project team to avoid nasty surprises like testing the wrong software, receiving uninstallable builds and reporting unreproducible defects against versions of code that do not exist anywhere but in the test environment.

5.5 DEFECT MANAGEMENT

Let's wind down the core, examinable material in this chapter on test management with an important subject: how we can document and manage the defects that occur during testing. One of the objectives of testing is to find defects, which reveal themselves as discrepancies between actual and expected outcomes. When we observe a defect, we need to log the details. Proper test management involves establishing appropriate actions to investigate and dispose of defects. This includes tracking defects from initial discovery and classification through to resolution, confirmation testing and ultimate disposition, with the defects following a clearly established set of rules and processes for defect management and classification. We'll look at what topics we should cover when reporting defects. At the end of this section, you will be ready to write a thorough defect report.

The Syllabus terms in this section are **defect management** and **defect report**.

5.5.1 Preparing a defect report

What are defect reports for?

When running a test, you might observe actual results that vary from expected results. This is not a bad thing—one of the major goals of testing is to find problems. Different organizations have different names to describe such situations. Commonly, they are called anomalies, incidents, bugs, defects, problems or issues.

To be precise, we sometimes draw a distinction between an incident or anomaly on the one hand and defects or bugs on the other. An incident or anomaly is any situation where the system exhibits questionable behaviour, and a defect is some problem, imperfection or deficiency in the item we are testing.

In addition to defects, causes of anomalies include misconfiguration or failure of the test environment, corrupted test data, bad tests, incorrect expected results and tester mistakes. In some cases, the policy is to classify as a defect anything that arises from a test design, the test environment or anything else which is under formal configuration management. Bear in mind the possibility that a questionable behaviour is not necessarily a true defect. We log these defects so that we have a record of what we observed and can follow it up and track what is done to correct it, whether or not it turns out to be a problem in the work product we are testing or something else. This is **defect management**.

> **Defect management**
> The process of recognizing, recording, classifying, investigating, resolving and disposing of defects.

While it is most common to find defect logging or defect reporting processes and tools in use during formal, independent testing, you can also log, report, track and manage defects found during development and reviews. In fact, this is a good idea, because it gives useful information on the extent to which early (and cheaper) defect detection and removal activities are happening.

Of course, we also need some way of reporting, tracking and managing defects that occur in the field or after deployment of the system. While many of these defects will be user error or some other behaviour not related to a problem in system behaviour, some percentage of defects do escape from quality assurance and testing activities. The defect detection percentage (DDP), which compares field defects with test defects, can be a useful metric of the effectiveness of the test process. We want the testers to find the defects before the users do, so the higher the DDP percentage, the more effective (at bug-catching) the testing was.

Here is an example of a DDP formula that would apply for calculating DDP for the last level of testing prior to release to the field:

$$DDP = \frac{defects\ (testers)}{defects\ (testers) + defects\ (field)}$$

It is most common to find defects reported against the code or the system itself. However, we have also seen cases where defects are reported against requirements and design specifications, user and operator guides and tests. Often, it aids the effectiveness and efficiency of reporting, tracking and managing defects when the defect tracking tool provides an ability to vary some of the information captured depending on what the defect was reported against.

In some projects, a very large number of defects are found. Even on smaller projects where 100 or fewer defects are found, you can easily lose track of them unless you have a process for reporting, classifying, assigning and managing the defects from discovery to final resolution.

Objectives for defect reports

A **defect report** contains a description of the misbehaviour that was observed and classification of that misbehaviour. As with any written communication, it helps to have clear goals in mind when writing. Typical objectives for such reports include the following:

Defect report
Documentation of the occurrence, nature, and status of a defect.

- To provide developers, managers and others with detailed information about the behaviour observed (the adverse event), that is, the defect. This will enable them to identify specific effects, to isolate the problem with a minimal reproducing test, and to correct the defect(s) as needed or to otherwise resolve the problem.

- To support test managers in the analysis of trends in aggregate defect data, either for understanding more about a particular set of problems or tests, or for understanding and reporting the overall level of system quality. This will enable them to track the quality of the work product and the impact on the testing. For example, if a lot of defects are being reported, the testers will do less testing, since their time is taken with writing defect reports. This also means that more confirmation tests will be needed.

- To enable defect reports to be analyzed over a project, and even across projects, to give information and ideas that can lead to development and test process improvements.

When writing a defect report, it helps to have the readers in mind, too. The developers need the information in the report to find and fix the defects. Before that happens, though, the defects should be reviewed and prioritized so that scarce testing and developer resources are spent fixing and confirmation testing the most important defects. Since some defects may be deferred—perhaps to be fixed later or perhaps, ultimately, not to be fixed at all—we should include work-arounds and other helpful information for help-desk or technical support teams. Finally, testers often need to know what their colleagues are finding so that they can watch for similar behaviour elsewhere and avoid trying to run tests that will be blocked.

How to write a good defect report: some tips

A good defect report is a technical document. In addition to being clear for its goals and audience, any good report grows out of a careful approach to researching and writing the report. We have some rules of thumb that can help you write a better defect report.

Use a careful, attentive approach to running your tests. You never know when you are going to find a problem. If you are pounding on the keyboard while gossiping with office mates or daydreaming about a movie you just saw, you might not notice strange behaviours. Even if you see the defect, how much do you really know about it? What can you write in your defect report?

Intermittent or sporadic symptoms are a fact of life for some defects, and it is always discouraging to have a defect report bounced back as irreproducible. It's a good idea to try to reproduce symptoms when you see them. We have found three times to be a good rule of thumb. If a defect has intermittent symptoms, we would still report it, but we would be sure to include as much information as possible, especially how many times we tried to reproduce it and how many times it did in fact occur.

You should also try to isolate the defect by making carefully chosen changes to the steps used to reproduce it. In isolating the defect, you help guide the developer to the problematic part of the system. You also increase your own knowledge of how the system works, and how it fails.

Think outside the box. Some test cases focus on boundary conditions, which may make it appear that a defect is not likely to happen frequently in practice. We have found that it's a good idea to look for more generalized conditions that cause the failure to occur, rather than simply relying on the test case. This helps prevent the infamous defect report rejoinder, 'No real user is ever going to do that'. It also cuts down on the number of duplicate reports that get filed.

As there is often a lot of testing going on with the system during a test period, there are lots of other test results available. Comparing an observed problem against other test results and known defects is a good way to find and document additional information that the developer is likely to find very useful. For example, you might check for similar symptoms observed with other defects, the same symptom observed with defects that were fixed in previous versions or similar (or different) results seen in tests that cover similar parts of the system.

Many readers of defect reports, managers especially, will need to understand the priority and severity of the defect. The impact of the problem on the users, customers and other stakeholders is important. Most defect tracking systems have a title or summary field and that field should mention the impact, too.

Choice of words definitely matters in defect reports. You should be clear and unambiguous. You should also be neutral, fact-focused and impartial, keeping in mind the testing-related interpersonal issues discussed in the additional material at the end of Chapter 1. Finally, keeping the report concise helps keep people's attention and avoids the problem of losing them in the details.

As a last rule of thumb for defect reports, we recommend that you use a review process for all reports filed. It works if you have the lead tester review reports and we have also allowed testers, at least experienced ones, to review other testers' reports. Reviews are proven quality assurance techniques and defect reports are important project deliverables.

What goes in a defect report?

A defect report describes some situation, behaviour or event that occurred during testing that was not as it should be, or that requires further investigation. In many cases, a defect report consists of one or two screens full of information gathered by a defect tracking tool and stored in a database.

A defect report filed during dynamic testing typically includes the following:

- An identifier.
- A title and short summary of the defect being reported.
- Date of the defect report, issuing organization, the date and time of the failure, the name of the tester, that is, the author of the defect report (and perhaps the reviewer of the test).
- Identification of the test item (configuration item being tested), the test environment and any additional information about the configuration of the software, system or environment.
- The development life cycle activity, activities or sprint in which the defect was observed.
- A description of the failure to enable reproduction and resolution, such as the steps to reproduce and the isolation steps tried.
- The expected and actual results of the test.
- The scope or degree of impact (severity) of the defect on the interests of stakeholder(s).
- Urgency/priority to fix.
- State of the defect report (for example, open, deferred, duplicate, waiting to be fixed, awaiting confirmation testing, reopened, closed).
- Conclusions, recommendations and approvals.
- Global issues, such as other areas that may be affected by a change resulting from the defect.
- Change history, such as the sequence of actions taken by project team members with respect to the defect, to isolate, repair and confirm it as fixed. These fields should mention the inputs given and outputs observed, the different ways you could—and could not—make the problem recur, and the impact.
- References, including the test case that revealed the problem, and references to specifications or other work products that provide information about correct behaviour.

Sometimes testers classify the scope, severity and priority of the defect, though sometimes managers or a bug triage committee handle that role.

If you are using a defect management tool, some of the information will be automatically generated, such as the time and date of the report, the ID number, the report author and the initial state (open).

An example defect report can be found in ISO/IEC/IEEE 29119-3 [2021], which uses the term incident report.

As the defect is managed to resolution, managers might assign a level of priority to the report. The change control board or bug triage committee might document the risks, costs, opportunities and benefits associated with fixing or not fixing the defect. The developer, when fixing the defect, can capture the root cause, the time or development stage where it was introduced and the time or testing activity that identified it, and remove it (which may be different from when it was identified).

After the defect has been resolved, managers, developers or others may want to capture conclusions and recommendations. Throughout the life cycle of the defect report, from discovery to resolution, the defect tracking system should allow each person who works on the defect report to enter status and history information.

What happens to defect reports after you file them?

As we mentioned earlier, defect reports are managed through a life cycle from discovery to resolution. The defect report life cycle is often shown as a state transition diagram (refer to Figure 5.6). While your defect tracking system may use a different life cycle, let's take this one as an example to illustrate how a defect report life cycle might work.

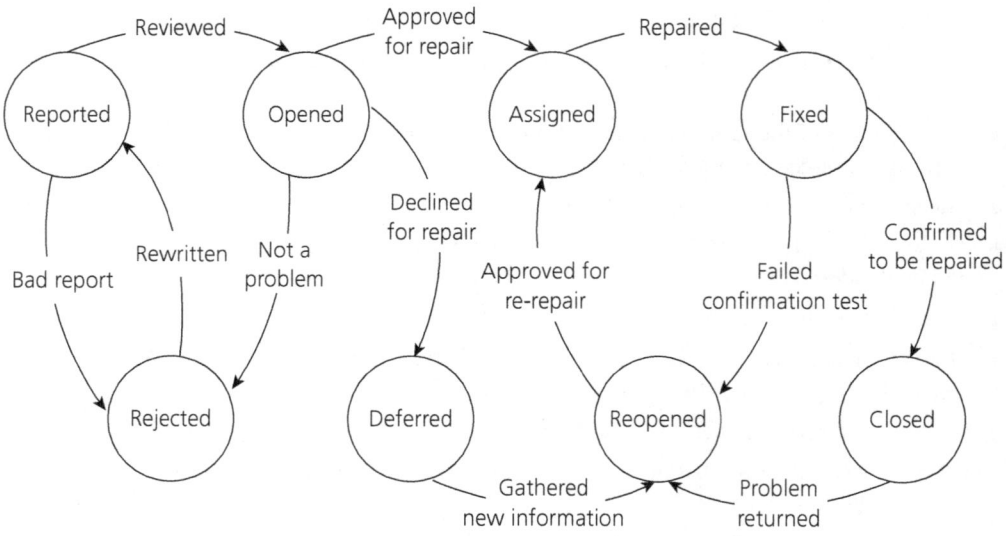

FIGURE 5.6   Defect report life cycle

In the defect report life cycle shown in Figure 5.6, all defect reports move through a series of clearly identified states after being reported. Some of these state transitions occur when a member of the project team completes some assigned task related to closing a defect report. Some of these state transitions occur when the project team decides not to repair a defect during this project, leading to the deferral of the defect report. Some of these state transitions occur when a defect report is poorly written or describes behaviour which is actually correct, leading to the rejection of that report.

Let's focus on the path taken by defect reports which are ultimately fixed. After a defect is reported, a peer tester or test manager reviews the report. If successful in the review, the defect report becomes opened, so now the project team must decide whether or not to repair the defect. If the defect is to be repaired, a developer is assigned to repair it.

Once the developer believes the repairs are complete, the defect report returns to the tester for confirmation testing. If the confirmation test fails, the defect report is re-opened and then re-assigned. Once the tester confirms a good repair, the defect report is closed. No further work remains to be done on this defect.

In any state other than rejected, deferred or closed, further work is required on the defect prior to the end of this project. In such a state, the defect report has a clearly identified owner. The owner is responsible for transitioning the defect into an allowed subsequent state. The arrows in the diagram show these allowed transitions.

In a rejected, deferred or closed state, the defect report will not be assigned to an owner. However, certain real world events can cause a defect report to change state even if no active work is occurring on the defect report. Examples include the recurrence of a failure associated with a closed defect report and the discovery of a more serious failure associated with a deferred defect report.

Ideally, only the owner can transition the defect report from the current state to the next state and ideally the owner can only transition the defect report to an allowed next state. Most defect tracking systems support and enforce the defect life cycle and life cycle rules. Good defect tracking systems allow you to customize the set of states, the owners and the transitions allowed to match your actual workflows. And, while a good defect tracking system is helpful, the actual defect workflow should be monitored and supported by project and company management.

CHAPTER REVIEW

Let's review what you have learned in this chapter.

From Section 5.1, you should now understand the fundamentals of test planning and estimation. You should know the reasons for writing test plans and be able to explain how test plans relate to projects, test levels, test objectives and test activities. You should be able to write a test execution schedule based on priority and dependencies. You should be able to explain the justification behind various entry and exit criteria that might relate to projects, test levels and/or test objectives. You should be able to distinguish the purpose and content of test plans from that of test design specifications, test cases and test procedures. You should know the factors that affect the effort involved in testing, including test strategies and approaches and how they affect testing, and be able to estimate test tasks. You should be able to explain how metrics, expertise and negotiation are used for estimating. You should know the Glossary terms **entry criteria**, **exit criteria**, **test approach**, **test plan**, **test planning**, **test pyramid** and **test quadrants**.

From Section 5.2, you should now be able to explain how risk is related to testing. You should know that a risk is a potential undesirable or negative outcome. You should know about likelihood and impact as factors that determine the importance of a risk. You should be able to compare and contrast risks to the product (and its quality) and risks to the project itself and know typical risks to the product and project. You should be able to describe how to use risk analysis and risk management for testing and test planning. You should understand how to control risk, including using testing to mitigate risks to product quality. You should know the Glossary terms **product risk**, **project risk**, **risk**, **risk analysis**, **risk assessment**, **risk control**, **risk identification**, **risk level**, **risk management**, **risk mitigation**, **risk monitoring** and **risk-based testing**.

From Section 5.3, you should be able to explain the essentials of test progress monitoring and control. You should know the common metrics that are captured, logged and used for monitoring, as well as ways to present these metrics. You should be able to explain a typical test progress report and test summary report, and how to use those to communicate to project and product stakeholders. You should know the Glossary terms **test control**, **test monitoring**, **test progress report** and **test completion report**.

From Section 5.4, you should now understand the basics of configuration management that relate to testing and how it helps us do our testing work better. You should know the Glossary term **configuration management**, though it is not specifically called out and thus is not examinable.

From Section 5.5, you should now understand defect logging and be able to use defect management on your projects. You should know the content of a defect report, be able to write a high quality report based on test results and manage that report through its life cycle. You should know the Glossary terms **defect management** and **defect report**.

SAMPLE EXAM QUESTIONS

Question 1 Which of the following factors is an influence on the test effort involved in most projects?

a. Geographical separation of tester and developers.

b. The departure of the test manager during the project.

c. The quality of the information used to develop the tests.

d. Unexpected long-term illness by a member of the project team.

Question 2 Consider the following exit criteria which might be found in a test plan:

I. No known customer-critical defects.
II. All interfaces between components tested.
III. 100% code coverage of all units.
IV. All specified requirements satisfied.
V. System functionality matches legacy system for all business rules.

Which of the following statements is true about whether these exit criteria belong in an acceptance test plan?

a. All statements belong in an acceptance test plan.

b. Only statement I belongs in an acceptance test plan.

c. Only statements I, II and V belong in an acceptance test plan.

d. Only statements I, IV and V belong in an acceptance test plan.

Question 3 According to the ISTQB Glossary, what is a test approach?

a. The manner of implementing testing tasks.

b. Documentation that expresses the generic requirements for testing one or more projects within an organization providing detail on how testing is to be performed.

c. Documentation describing the test objectives to be achieved and the means and the schedule for achieving them, organized to coordinate testing activities.

d. The set of interrelated activities comprising of test planning, test monitoring and control, test analysis, test design, test implementation, test execution and test completion.

Question 4 Which of the following metrics would be most useful to monitor during test execution?

a. Percentage of test cases written.

b. Number of test environments remaining to be configured.

c. Number of defects found and fixed.

d. Percentage of requirements for which a test has been written.

Question 5 What would appear in a test progress report that would not be included in a test completion report?

a. Summary of testing performed.

b. The status of the test activities against the test plan.

c. Factors that have blocked or continue to block progress.

d. Reusable test products produced.

Question 6 In a defect report, the tester makes the following statement: 'The payment processing subsystem fails to accept payments from American Express cardholders, which is considered a must-work feature for this release.' This statement is likely to be found in which of the following sections?

a. Scope or degree of impact of the defect.

b. Identification of the test item.

c. Summary of the defect.

d. Description of the defect.

Question 7 During an early period of test execution, a defect is located, resolved and passed its confirmation test, but is seen again later during subsequent test execution. Which of the following is a testing-related aspect of configuration management that is most likely to have broken down?

a. Traceability.

b. Configuration item identification.

c. Configuration control.

d. Test documentation management.

Question 8   You are working as a tester on a project to develop a point-of-sales system for grocery stores and other similar retail outlets. Which of the following is a product risk for such a project?

a. The arrival of a more reliable competing product on the market.

b. Delivery of an incomplete test release to the first cycle of system test.

c. An excessively high number of defect fixes fail during confirmation testing.

d. Failure to accept allowed credit cards.

Question 9   What is a risk?

a. A bad thing that has happened to the product or the project.

b. A bad thing that might happen.

c. Something that costs a lot to put right.

d. Something that may happen in the future.

Question 10   You are writing a test plan and are currently completing a section on risk. Which of the following is most likely to be listed as a project risk?

a. Unexpected illness of a key team member.

b. Excessively slow transaction processing time.

c. Data corruption under network congestion.

d. Failure to handle a key use case.

Question 11   You and the project stakeholders develop a list of product risks and project risks during the planning stage of a project. What else should you do with those lists of risks during test planning?

a. Determine the extent of testing required for the product risks and the mitigation and contingency actions required for the project risks.

b. Obtain the resources needed to completely cover each product risk with tests and transfer responsibility for the project risks to the project manager.

c. Execute sufficient tests for the product risks, based on the likelihood and impact of each product risk and execute mitigation actions for all project risks.

d. No further risk management action is required at the test planning stage.

Question 12   The following are ways to estimate test effort:

1. Burndown charts.

2. Wideband Delphi.

3. Expert opinion.

4. Previous effort for this team.

5. Planning poker.

6. Similar project data.

Which of these are an example of a metrics-based technique?

a. 3, 4 and 5.

b. 1, 4 and 6.

c. 2, 3 and 4.

d. 1, 2 and 5.

Question 13   In a defect report, the tester makes the following statement: 'At this point, I expect to receive an error message explaining the rejection of this invalid input and asking me to enter a valid input. Instead, the system accepts the input, displays an hourglass for between one and five seconds and finally terminates abnormally, giving the message, 'Unexpected data type: 15. Click to continue'. This statement is likely to be found in which of the following sections of a defect report?

a. Summary.

b. Impact.

c. Item pass/fail criteria.

d. Defect description.

Question 14   There are five tests to be executed. A high-priority test has been scheduled as the third in the sequence of five, rather than the first. Which of the following would NOT be a good reason for this?

a. The high priority test has a logical dependency on the two tests run before it.

b. All tests are the same priority, and this is the most efficient sequence.

c. The tests scheduled before it are more likely to pass so this will take less time and give feedback more quickly.

d. The first three tests have the same priority, so it does not matter what order they are run in.

Question 15 Exit criteria have been agreed at the start of the project, but it is now clear that they will not all be met before the release date next week. What should the test manager NOT say to those who want to release next week anyway?

a. These exit criteria were agreed for a very good reason, to avoid this situation. It is essential that all exit criteria are met before we release.

b. It is not my role to make that decision, so if you go ahead and release next week anyway, I will carefully monitor the effects and will report the results to high-level management.

c. Have all project stakeholders, business owners and the product owner been informed of the risks of releasing next week, and do they all understand and accept them? (Sign on this paper stating this.)

d. What are the most critical exit criteria, and can we get agreement from all stakeholders about which are absolutely necessary, and which could be compromised this time?

Question 16 Which of the following is a way in which a tester adds value during *iteration* planning?

a. By helping to define the initial product backlog.

b. By executing non-functional tests.

c. By performing a detailed risk analysis on the selected user stories.

d. By defining the test approach across all iterations.

Question 17 You find the following statement in a test plan:
All test environment tests have passed.
    Which section of the test plan would you expect this to be in?

a. Entry criteria.

b. Definition of done.

c. Exit criteria.

d. Risk register.

Question 18 Consider a test case that took 2.5 hours, 3 hours and 2 hours to run the last three times it was executed. Which of the following statements is true?

a. Ratio estimation says we should estimate 2.5 hours for the next execution.

b. Three-point estimation says that the measurement error is 1 hour.

c. Extrapolation estimation says we should estimate 1.5 hours for the next execution.

d. Extrapolation estimation says we should estimate 2.5 hours for the next execution.

Question 19 The following table shows test procedures, dependencies and priorities (larger number means higher priority) for an e-commerce system.

Test Procedure ID	Summary	Dependency	Priority
A	Locate an item in inventory	none	3
B	Place item in shopping cart	A	4
C	Checkout and confirm item is in fulfilment process for buyer	B	5
D	Save item for later purchase	B	2
E	Purchase saved item and confirm item is in fulfilment process for buyer	D	2

Which of the following is the optimal test execution schedule?

a. A, B, A, B, C, D, E

b. B, A, D, E, B, A, C

c. A, B, D, E, A, B, C

d. A, B, C, A, B, D, E

Question 20   Which of the following statements is accurate about the testing pyramid?

a. Groups test types and levels.

b. Guides selection of test types.

c. Emphasizes upper-level tests.

d. Emphasizes lower-level tests.

Question 21 Which of the following statements is true about user acceptance testing in the Testing Quadrants model?

a. User acceptance testing is typically automated.

b. User acceptance testing occurs in continuous integration.

c. User acceptance testing verifies the acceptance criteria.

d. User acceptance testing validates the product.

Question 22 Which of the following does NOT mitigate product risks?

a. Assign appropriate testers.

b. Use performance tests.

c. Set up a technical support group.

d. Run regression tests after changes.

Question 23 A tester submits a test progress report that contains the content described within ISO 29119-3. This is most likely an example of which kind of test status communication?

a. Verbal communication.

b. Formal reporting.

c. CI/CD dashboards.

d. Email communication.

EXERCISES

Test case prioritization exercise

You have been asked to develop a test execution schedule for the second release of a customer data management application. The first release only allowed basic customer records to be added. This second one contains five new features and two bug fixes, together with the new features' priorities (1 is highest) and the defect severity levels, in the table below.

TABLE 5.3   Exercise: Test execution schedule

Priority	Feature/Fix	Test order
1	Delete inactive customer record	
2	De-activate customer	
3	Edit extra data items	
4	Edit basic data	
5	Link customer records (for example in the same company)	
5	Provide extra data items when record added	

What would be the BEST test execution sequence for this release?

Defect report exercise

Assume you are testing a maintenance release which will add a new supported model to a browser-based application that allows car dealers to order custom-configured cars from the maker. You are working with a selected number of dealers who are performing beta testing. When they find problems, they send you an email describing the failure. You use that email to create a defect report in your defect management system.

You receive the following email from one of the dealers:

I entered an order for a Racinax 917X in midnight violet. During the upload, I got an 'unexpected return value' error message. I then checked the order, and it was in the system. I think the internet connection might have gone down for a moment as we found out later that there was a power surge elsewhere in the building at around that time.

Write a defect report based on this email and note any elements of such a report which might not be available based on this email alone.

Note that this scenario, including the name of the car model, is entirely fictitious.

Test estimation exercise

Examining the history of a given test case for the last few times it was run, we see that that effort required, in person-hours to the nearest half-hour, was:

$$10.0, 6.5, 4.5, 4.0, 5.5, 6.0, 5.5, 4.0, 12.0, 4.5$$

Applying a metrics-based approach, what would you expect the next test execution effort figure to be?

EXERCISE SOLUTIONS

Test case prioritization exercise

Below we show the solution, with the tests listed in the test order. The justifications are shown below the table.

TABLE 5.4 Solution: Test execution schedule

Priority	Feature/Fix	Test order
2	De-activate customer	1
1	Delete inactive customer record	2
5	Provide extra data items when record added	3
3	Edit extra data items	4
4	Edit basic data	5
5	Link customer records (for example in the same company)	6

The highest priority is to Delete an inactive customer, but first we must have an inactive customer, so we run De-activate customer first. This is a functional dependency.

The next highest priority is to Edit extra data items, but again, we need to have some extra data items in order to do this, another functional dependency.

The final two are independent, so the priority 4 test is run before the priority 5 test.

Defect report exercise

Here is an example of a defect report:

Test defect report identifier: This would be assigned by the defect tracking tool.

Summary: System returns confusing error message if internet connectivity is interrupted.

Defect description:
   Steps to reproduce

1. Entered an order for a Racinax 917X in midnight violet.
2. During the order upload, the system displayed an 'unexpected return value' error message.
3. Verified that, in spite of the error message, the order was in the system.

Suspected cause Given that a brief interruption in the internet connection may have occurred during order transmission due to the power surge, the suspected cause of the failure is a lack of robust handling of slow or unreliable internet connections.

Impact assessment The message in step 2 is not helpful to the user, because it gives the user no clues about what to do next. A user encountering the message described in step 2 might decide to re-enter the order, which in this case would result in a redundant order.

Some car dealerships will have unreliable internet connections, with the frequency of connection loss depending on location, wireless infrastructure available in the dealership, type of internet connection hardware used in the dealers' computers, and other such factors. Therefore we can expect that some number of defects such as this will occur in widespread use. (Indeed, this beta test defect proves that this is a likely event in the real world.) Since a careless or rushed dealer might decide to resubmit the order based on the error message, this will result in redundant orders, which will cause significant loss of profitability for the dealerships and the salespeople themselves.

This report addresses the inputs, the expected results, the actual results, the difference between the expected and actual results and the procedure steps in which the tester identified the unexpected results. Here is some additional information needed to improve the report:

- The failure needs to be reproduced at least two more times. By doing the failure replication in the test environment, the tester will be able to control whether or not the internet connection goes down during the order and for how long the connection is down.

- Is the problem in any way specific to the Racinax 917X model and/or the colour? Isolation of the failure is needed, and then the report can be updated.

- Is the problem in any way specific to power surges? We would guess not, but checking with different types of connections (during the isolation and replication discussed above) would make sense. If it proves independent of the type of connection, we would know that the problem is more general than just for power surges.

- The defect report should also include information about the date and time of the test, the beta test environment and the name of the beta tester and the tester entering the defect.

Test estimation solution

Purely applying extrapolation to all ten past executions, we find the average is 6.5 (to the nearest half-hour). However, note that we have two outliers in the history: 10.0 and 12.0. Outliers have a distorting effect on averages. So, a more accurate figure might be 5.0, which is the average if we remove the outliers, or we could use the median value, which is 5.5.

CHAPTER SIX

Test tools

Y ou may be wishing that you had a magic tool that would automate all the testing for you. If so, you will be disappointed. However, there are a number of very useful tools that can bring significant benefits. In this chapter, we will see that there is tool support for many different aspects of software testing. We will see that success with tools is not guaranteed, even if an appropriate tool is acquired. There are also risks in using tools.

Note that in this chapter we have included a lot of additional useful information about tools and test automation. If you are just studying for the exam, focus on the Syllabus and the Learning Objectives; if you want to improve your automation, take note of all the advice here.

6.1 TOOL SUPPORT FOR TESTING

> **SYLLABUS LEARNING OBJECTIVE FOR 6.1 TOOL SUPPORT FOR TESTING**
>
> **FL-6.1.1**   **Explain how different types of test tools support testing (K2)**

In this section, we will describe the various tool types in terms of the testing activities that they support and facilitate. We will not be going into lots of detail. The reason for this is that, in general, the types of tools will be fairly stable over a longer period, even though there will be new vendors in the market, new and improved tools, and even new types of tools in the coming years.

We will not mention any specific open source or commercial tools in this chapter. If we did, this book would date very quickly! Tool vendors are acquired by other vendors, change their names and change the names of the tools quite frequently, so we will not mention the names of any tools or vendors.

Management tools

Management tools can help to increase the efficiency of the test process in a number of different aspects. Managing the software development life cycle, including test activities, can help managers keep track of the tasks and progress of software development projects; this type of tool may also be called an Application Lifecycle

Management (ALM) tool. A test management tool may also keep track of the tests, including traceability to requirements or other tests. Requirements management tools store and manage system requirements in a central place. Defect management tools (or bug tracking tools) keep track of defects reported, their status and progress from initial report to resolution. Configuration management tools keep track of different versions of testware as well as software.

Let's look at these types of tools in more detail.

What does test management mean? It could be the management of tests, or it could be managing the testing process. The tools in this broad category provide support for either or both of these. The management of testing applies over the whole of the software development life cycle, so a test management tool could be among the first to be used in a project. A test management tool may also manage the tests, which would begin early in the project and would then continue to be used throughout the project and also after the system had been released. In practice, test management tools are typically used by specialist testers or test managers at system or acceptance test level.

Test management tools

Test management tools provide features that cover both the management of testing, such as progress reports and keeping track of testing activities, and the management of testware, such as logging of test results and keeping track of test environments needed for tests. There are some tools that provide support for only one activity (for example, defect management tools); other tools or tool suites may provide support for all test management activities.

ALM tools manage not only testing but also development and deployment. With a focus on communication, collaboration and task tracking, they are popular in Agile development.

The information from test management tools (and ALM tools) can be used to monitor the testing process and decide what actions to take (test control), as described in Chapter 5. The tool also gives information about the component or system being tested (the test object). Test management tools help to gather, organize and communicate information about the testing on a project.

Requirements management tools

Are requirements management tools really testing tools? Some people may say they are not, but they do provide some features that are very helpful to testing. Because tests are based on requirements, the better the quality of the requirements, the easier it will be to write tests from them. It is also important to be able to trace tests to requirements and requirements to tests, as we saw in Chapter 2. Note that requirements means any source of what the system should do, such as user stories.

Defect management tools

This type of tool is also known as a defect tracking tool, an incident management tool, a bug tracking tool or a bug management tool. However not all the things tracked are actually defects or bugs. There may also be perceived problems, anomalies (that aren't necessarily defects) or enhancement requests. This is why they are sometimes referred to as incidents and incident management tools. Also, what is normally recorded is information about the failure (not the defect) that was generated during testing. Information about the defect that caused that failure would come to light when someone (for example, a developer) begins to investigate the failure.

Defect reports go through a number of stages, from initial identification and recording of the details, through analysis, classification, assignment for fixing, fixed, re-tested and closed, as described in Chapter 5. Defect management tools make it much easier to keep track of the defects and their status over time.

Configuration management tools

Assume that a test group began testing the software, expecting to find the usual fairly high number of problems. But to their surprise, the software seemed to be much better than usual this time. Very few defects were found. Before they celebrated the great quality of this release, they just made an additional check to see if they had the right version and discovered that they were actually testing the version from two months earlier (which had been debugged) with the tests for that earlier version. It was nice to know that this was still OK, but they were not actually testing what they thought they were testing or what they should have been testing.

Configuration management tools are not strictly testing tools either, but good configuration management is critical for controlled testing, as was described in Chapter 5. We need to know exactly what it is that we are supposed to test, such as the exact version of all the things that belong in a system. It is possible to perform configuration management activities without the use of tools, but the tools make life a lot easier, especially in complex environments. The same tool may be used for testware as well as for software items. Testware also has different versions and is changed over time. It is important to run the correct version of the tests as well, as our example shows.

Static testing tools

The tools described in this section support the testing activities described in Chapter 3. There are two types of tools: tools that support the review process and static analysis tools. Review process support tools keep track of who is reviewing, progress of the review, defects found and other aspects of good review practices. Static analysis tools examine code without executing it and can find anomalies such as dead code or broken links. Static analysis tools are often used by developers. Static analysis tools also play a key role in automated build and delivery processes associated with Agile development and DevOps.

Let's look at these tools in more detail.

Tools that support reviews

The value of different types of review was discussed in Chapter 3. For a very informal review, where one person looks at another's work product and gives a few comments about it, a tool such as this might just get in the way. However, when the review process is more formal, when many people are involved, or when the people involved are in different geographical locations, then tool support becomes far more beneficial.

It is possible to keep track of all the information for a review process using spreadsheets and text documents, but a review tool that is designed for the purpose is likely to do a better job. For example, one thing that should be monitored for each review is that the reviewers have not gone over the work product too quickly, that is, that the checking rate (number of pages checked per hour) was close to that recommended for that review cycle. A review tool could automatically calculate the checking rate and flag exceptions. The review tools can normally be tailored for the particular review process or type of review being done.

Static analysis tools

Static analysis tools are normally used by developers as part of the development and component testing process. The key aspect is that the code (or other artefact) is not executed or run. Of course, the tool itself is executed, but the source code we are interested in is the input data to the tool.

Static analysis tools are an extension of compiler technology, in fact some compilers do offer static analysis features. It's worth checking what is available from existing compilers or development environments before looking at acquiring a more sophisticated static analysis tool.

Static analysis can also be carried out on things other than software code, for example static analysis of requirements or static analysis of websites to assess for proper use of accessibility tags or the following of HTML standards.

Static analysis tools for code can help the developers to understand the structure of the code and can also be used to enforce coding standards.

Test design and implementation tools

The tools described in this section support the testing activities described in Chapter 4. These tools help to create maintainable test designs, test cases, test procedures and test data. This type of tool includes Model-Based Testing tools, which can generate tests from the models (e.g. state transition models). Test data preparation tools can help to manipulate data, for example making personal data anonymous.

Test design test tools

Test design tools help to construct test cases, or at least test inputs (part of a test case). If an automated test oracle is available, then the tool can also construct the expected result, so it can actually generate test cases, rather than just test inputs.

Tests that are designed using a tool need to have a starting point to derive the tests from. This could be requirements from a requirements management tool, which may identify valid and invalid boundary values for an input field, for example. Tests can be derived from elements on a screen, to ensure that each element is exercised by a test, for example buttons, input fields or pull-down lists. Tests can also be derived from a model of the system, as in model-based testing (discussed shortly). Coverage tools may identify the tests needed to extend white-box coverage.

There are two problems with test design tools. First, although it is relatively easy to generate test inputs, automatically generating the correct expected result is more challenging. Sometimes an oracle is available (for example, valid boundary conditions), other times only a partial oracle (generating some aspect but not all detail of an expected result), and most often no oracle other than 'The system is still running'.

The second problem is having too many tests and needing to find a way of identifying the most important tests to run. Cutting down an unmanageable number of tests can be done by risk analysis (refer to Chapter 5).

Model-based testing tools

Tools that generate test inputs or test cases from stored information that represents some kind of model of the system or software are called Model-Based Testing (MBT) tools. Such tools may be based on a state diagram, to implement state transition testing, for example.

MBT tools work by generating tests (inputs or test cases) automatically based on what is stored about the model used to describe the system behaviour. If the system is changed, only the model is updated, and the new tests are then generated automatically.

More information about MBT is available in the ISTQB Specialist Level Model-Based Tester Syllabus and qualification.

Test data preparation tools

In test implementation, setting up test data can be a significant effort, especially if an extensive range or volume of data is needed for testing. Test data preparation tools help in this area. They may be used by developers, but they may also be used during system or acceptance testing. They are particularly useful for performance and reliability testing, where a large amount of realistic data is needed.

The most sophisticated tools can deal with a range of files and database formats. These tools are also useful for anonymizing data to conform to data protection rules.

Test execution and coverage tools

When people think of a testing tool, it is usually a test execution tool that they have in mind, a tool that can run tests. This is a very popular type of tool especially for regression tests, where a large number of tests need to be repeated often; in fact tools are ideal for this. To work directly with this type of tool, software development expertise is needed, as the tools use scripting languages. Refer to Fewster and Graham [1999], Buwalda *et al.* [2001], Graham and Fewster [2012] and Gamba and Graham [2018].

Coverage tools measure how much has been tested by a specific set of tests. Typically, developers would use coverage tools to measure code coverage such as statement or branch coverage (refer to Section 4.3).

Let's look at these two tool types in more detail.

Test execution tools

This type of tool is also referred to as a test running tool. Some tools of this type offer a way to get started by capturing or recording manual tests; hence they are also known as capture/playback tools, capture/replay tools or record/playback tools. The analogy is with recording a television programme and playing it back. However, the tests are not something which is played back just for someone to watch!

Test execution tools use a scripting language to drive the tool. The scripting language is actually a programming language. Testers who wish to use a test execution tool directly may need to use programming skills to create and modify the scripts, although newer tools may provide a higher-level interface for testers (low-code or no-code). The advantages of programmable scripting: tests can repeat actions (in loops) for different data values (test inputs), they can take different routes depending on the outcome of a test (for example, if a test fails, go to a different set of tests) and they can be called from other scripts giving modular structure to the tests.

When people first encounter a test execution tool, they tend to use it to capture (or record) tests, which sounds really good when you first hear about it. However, the approach breaks down when you try to replay the captured tests. This approach does not scale up for large numbers of tests. The main reason for this is that a captured script is very difficult to maintain because:

● It is closely tied to the flow and interface presented by the GUI (graphical user interface).

● It may rely on the circumstances, state and context of the system at the time the script was recorded. For example, a script will capture a new order number assigned by the system when a test is recorded. When that test is played back, the system will assign a different order number and reject subsequent requests that contain the previously captured order number.

- The test input information is hard-coded, that is, it is embedded in the individual script for each test.

Any of these things can be overcome by modifying the scripts, but then we are no longer just recording and playing back! If it takes more time to update a captured test than it would take to run the same test again manually, the scripts tend to be abandoned and the tool is abandoned.

There are better ways to use test execution tools to make them work well and actually deliver the benefits of unattended automated test running. There are at least five levels of scripting and also different comparison techniques. Data-driven testing is an advance over captured scripts, but keyword-driven testing gives significantly more benefits.

There are many different ways to use a test execution tool and the tools themselves are continuing to gain new useful features. Test execution tools are often used for regression testing, where tests are run that have been run before, such as in continuous integration. One of the most significant benefits of using this type of tool is that whenever an existing system is changed (for example, for a defect fix or an enhancement), the tests that were run earlier can be run again, to make sure that the changes have not disturbed the existing system.

Coverage tools

How thoroughly have you tested? Coverage tools can help answer this question. Requirements coverage tools measure how many requirements have been exercised by a set of tests. Code coverage tools, normally used by developers, measure how many code elements, such as statements or branches, have been exercised by a set of tests.

A coverage tool first identifies the elements or coverage items that can be counted, and where the tool can identify when a test has exercised that coverage item. At component testing level, the coverage items could be lines of code or code statements or decision outcomes (for example, the True or False exit from an IF statement). At component integration level, the coverage item may be a call to a function or module. At system or acceptance level, the coverage item may be a user story, a requirement, a function or a feature.

The coverage tool counts the number of coverage items that have been executed by the test suite and reports the percentage of coverage items that have been exercised and may also identify the items that have not yet been exercised (that is, not yet tested). Additional tests can then be run to increase coverage.

Note that the coverage tools only measure the coverage of the items that they can identify. Just because your tests have achieved 100% coverage does not mean that your software is 100% tested! Refer also to Chapter 4 Section 4.3.

Non-functional testing tools

The tools described in this section support testing that is difficult or impossible to perform manually. Whereas functional tests are looking at what results the system produces, non-functional tests look at how well the system behaves This type of tool includes performance testing tools, monitoring tools, usability testing tools and security testing tools, which we will describe in more detail.

Performance testing tools

Performance testing tools are concerned with testing at system level to see whether or not the system will stand up to a high volume of usage. A load test checks that the system can cope with its expected number of transactions. A volume test checks that the system can cope with a large amount of data, for example many fields in a record, many records in a file, etc. A stress test is one that goes beyond the normal expected usage of the system (to see what would happen outside its design expectations), with respect to load or volume.

In performance testing, many test inputs may be sent to the software or system where the individual results may not be checked in detail. The purpose of the test is to measure characteristics, such as response times, throughput or the mean time between failures (for reliability testing). Load generation can simulate multiple users or high volumes of input data. More sophisticated tools can generate different user profiles, different types of activity, timing delays and other parameters. Adequately replicating the end-user environments or user profiles is usually key to realistic results. Performance testing tools normally provide reports based on test logs and graphs of load against response times.

Analyzing the output of a performance testing tool is not always straightforward and it requires time and expertise. If the performance is not up to the standard expected, then some analysis needs to be performed to see where the problem is and to know what can be done to improve the performance.

Monitoring tools

Monitoring tools are used to continuously keep track of the status of the system in use, in order to have the earliest warning of problems and to improve service. There are monitoring tools for servers, networks, databases, security, performance, website and internet usage, and applications.

Monitoring tools may help to identify performance problems and network problems, and give information about the use of the system, such as the number of users.

Usability testing tools

Tools to support usability testing can help to assess user experience in using the system, for example surveys after using a website or mobile app. Tools can check to make sure there are no broken links on a web page. Tools can also monitor usage, such as which links are clicked most often. Video recorders and screen capture utilities can also be used to help assess usability. Major companies that sell to the public may also have a usability lab, where beta versions of software are used by volunteers from the target market; this could include video, analysis of key presses or even eye movement tracking.

Security testing

There are a number of tools that protect systems from external attack, for example firewalls, which are important for any system.

Security testing tools can be used to test security by trying to break into a system, whether or not it is protected by a security tool. The attacks may focus on the network, the support software, the application code or the underlying database.

Security testing tools can provide support in identifying viruses, detecting intrusions such as denial of service attacks, simulating various types of external attacks, probing for open ports or other externally visible points of attack, and identifying weaknesses in password files and passwords. In addition, these tools can perform security checks during operation, for example checking the integrity of files, intrusion detection and the results of test attacks.

DevOps tools

DevOps tools support the delivery pipeline, workflow tracking, automated build processes and CI/CD (Continuous Integration/Continuous Development/Delivery). This type of tool may include unit test framework tools, test harnesses, mock objects, static analysis tools and continuous integration tools.

Incremental and iterative development life cycles require frequent builds, and DevOps and continuous integration tools are an essential part of the Agile toolkit. Continuous integration is the practice of integrating new or changed code with the existing code repository very frequently, at least daily but sometimes dozens of times a day or more. Unit tests are most often automatically run when a new build is made (when a developer commits their change), so any defects found can be corrected immediately. The result of the integration is a system that is tested and could in principle be deployed to users at any time.

If the deployment is automated, then it is called continuous delivery, but there might still be some final human approval before it is released. If deployment is automatic (with no final human approval), then it is called continuous deployment. Continuous integration, delivery and deployment are the basis for DevOps, where code changes are delivered to users in an automated delivery pipeline.

These tools are continuous because the integration of the code components happens so often and is automatic (without human intervention), whenever the new build is triggered by a developer. This type of tool is included here because it includes tests which are managed in the continuous integration tool.

Collaboration tools

Collaboration tools are used to help facilitate good communication within the team. Good communication is essential for efficient working and applies equally to testers, developers, managers and others. Collaboration tools include shared internet resources such as virtual white-boards, the sharing of files and, especially since the COVID-19 pandemic, virtual meetings.

Tools supporting scalability and deployment standardization

This type of tool includes virtual machines and containerization tools. A virtual machine simulates a computer or an operating system so that tests can be run in the simulated environment rather than using actual computers. This is important where the ultimate hardware may not be available yet, but it also allows many tests to be run in parallel on different virtual machines (sometimes in the cloud). Containerization is the simulation of smaller parts of the system in self-contained units (containers). Dependencies are minimized so that each container can run independently.

Any other tool that assists in testing

In addition to the tools that support specific testing activities, any other tool that helps in testing can be regarded as a testing tool. For example, if testers use a spreadsheet for test cases, then that spreadsheet is a type of testing tool too.

Other types of tools that can help testing include data quality assessment tools, data conversion and migration tools, accessibility testing tools and localization testing tools, as well as general purpose applications such as spreadsheets.

6.2 BENEFITS AND RISKS OF TEST AUTOMATION

> **SYLLABUS LEARNING OBJECTIVE FOR 6.2 BENEFITS AND RISKS OF TEST AUTOMATION (K1)**
>
> FL-6.2.1   Recall the benefits and risks of test automation (K1)

Test automation The use of software to perform or support test activities.

The only Syllabus term for this chapter is **test automation**. You will find this term and its definition, taken from the ISTQB Glossary, in this chapter, in the Glossary and online.

The reason for acquiring tools to support testing is to gain benefits, by using software to do certain tasks that are better done by a computer than by a person.

The term **test automation** is most often used to refer to tools that execute tests and compare results, that is, test execution tools. This type of tool is one which can provide great benefits, but there are also significant risks. Test automation is also much broader than just supporting execution; tools are available to support many aspects of testing.

Potential benefits of using tools

There are many benefits that can be gained by using tools to support testing, whatever the specific type of tool. These are discussed next.

Time saved by reducing repetitive manual work

Repetitive work is tedious to do manually. People become bored and make mistakes when doing the same task over and over. Examples of this type of repetitive work include running regression tests, entering the same test data over and over again (both of which can be done by a test execution tool), comparing actual with expected results, checking against coding standards (which can be done by a static analysis tool) or running a large number of tests through the system in a short time (which can be done by a performance testing tool). Tools can also help to set up or tear down test environments. Using tools to help with repetitive work can save time (as well as frustration).

Prevention of simple human errors through greater consistency and repeatability

People tend to do the same task in a slightly different way even when they think they are repeating something exactly. A tool will exactly reproduce what it did before, so each time it is run, the result is consistent. Examples of where this aspect is beneficial include checking to confirm the correctness of a fix to a defect (which can be done by a debugging tool or test execution tool), entering test inputs (which can be done by a test execution tool), creating test data in a systematic manner and generating tests from requirements (which can be done by a test design tool, MBT tool or possibly a requirements management tool). The tool will run the same tests in the same order (unless told to do something different).

More objective assessment

If a person calculates a value from the software or defect reports, they may inadvertently omit something, or their own subjective prejudices may lead them to interpret that data incorrectly. Using a tool means that subjective bias is removed and the assessment is more repeatable and consistently calculated. Examples include assessing the structure of a component (which can be done by a static analysis tool), measuring coverage of the test item by a set of tests (coverage measurement tool), assessing system behaviour (monitoring tools) and assessing defect statistics (test management tool or defect management tool). Tools also may provide measurements that are too complex for humans to derive manually or take measurements more precisely than a human could.

Easier access to information about testing

Having lots of data does not mean that information is communicated. Information presented visually is much easier for the human mind to take in and interpret. For example, a chart or graph is a better way to show information than a long list of numbers. This is why charts and graphs in spreadsheets are so useful. Special-purpose tools provide these features directly for the information they process. Examples include statistics and graphs about test progress (test execution tool or test management tool), defect rates (defect management or test management tool) and performance (performance testing tool).

In addition to these general benefits, each type of tool has specific benefits relating to the aspect of testing that the particular tool supports. These benefits are normally prominently featured in the information available for the type of tool. It is worth investigating a number of different tools to get a general view of the benefits.

Reduced test execution times

Those tests that are run by a test execution tool take a fraction of the time that would be needed to run those same tests manually. If tests are run more quickly, defects can be identified earlier, giving more time to fix them. By providing a quick check on quality, we are shortening the feedback loop, so the code that has failed will be fresher in the developer's mind. Faster execution can also result in faster time to market (assuming the defects are not too extensive and are fixed quickly). Note that this benefit needs to be balanced with extra time that may be needed for other (non-execution) test activities, such as analysis of defects identified by the tool.

More time for testers to design tests

This is one of the major benefits of faster test execution. Because the testers' time is not taken up with entering test inputs, executing tests and evaluating results, they can spend that time devising new tests that probe the software more deeply and are more effective. It frees them to do more exploratory testing, which is much more effective at finding defects than automated tests due to Principle 5: Tests wear out (also called the Pesticide Paradox) as described in Section 1.3. At its best, test automation sets the testers free to do better testing.

Risks of using tools

Although there are significant benefits that can be achieved using tools to support testing activities, there are many organizations that have not achieved the benefits they expected.

Simply purchasing a tool is no guarantee of achieving benefits, just as buying membership in a gym does not guarantee that you will be fitter. Each type of tool requires investment of effort and time in order to achieve the potential benefits.

There are many risks that are present when tool support for testing is introduced and used, whatever the specific type of tool. We look at these next.

Unrealistic expectations for the benefits of test automation

Unrealistic expectations may be one of the greatest risks to success with tools. The tools are only software, and we all know that there are many problems with any kind of software! It is important to have clear objectives for what the tool can do and that those objectives are realistic. Unrealistic expectations may be both for functionality and for ease of use.

Inaccurate estimation of time, cost and effort

We think that 'inaccurate estimation' actually means under-estimating! Do you know of any large government project where the costs were over-estimated? Neither do we. The same is true within an organization introducing a test automation tool. Introducing something new into an organization is seldom straightforward. Having acquired a tool, you will want to move from downloading it to having a number of people being able to use the tool in a way that will bring benefits. There will be technical problems to overcome, but there will also be resistance from other people—both need to be addressed in order to succeed in introducing a tool. Often forgotten are the time, cost and effort or external expertise required to help get things onto the right track, and for training, not just in the use of the tool itself but also internal training after a pilot project so that everyone is using the tool and internal conventions are agreed in a consistent way for maximum benefit. Simply acquiring a tool and getting it started is not enough either. When a tool is introduced, it will affect the way testing is done, so test processes and test documentation will need to change.

Think back to the last time you did something new for the very first time (learning to drive, riding a bike, skiing). Your first attempts were unlikely to be very good, but with more experience you became much better. Using a testing tool for the first time will not be your best use of the tool either. It takes time to develop ways of using the tool to achieve what's possible. In the excitement of acquiring a new tool, it is easy to forget about things like the need for changes in the test process as the tool is implemented. These changes should be planned from the start.

But it doesn't stop there. It is critical for continuing success that there is continuous improvement in the way the tool is used. This may involve re-factoring automation assets from time to time or changing the way those assets are organized. Fortunately, there are some short cuts (for example, reading books and articles about other people's experiences and learning from them). Refer also to the additional material on effective use of tools for more detail on introducing a tool into an organization.

Insufficient planning for maintenance of the assets that the tool produces and uses is a strong contributor to tools that end up not being used. Although particularly relevant for test execution tools, planning for maintenance is also a factor with other types of tool. Even an open source tool needs investment.

Relying on a tool too much

Tools are definitely not magic! They can do very well what they have been designed to do (at least a good quality tool can), but they cannot do everything. A tool can certainly help, but it does not replace the intelligence needed to know how best to use it and how to evaluate current and future uses of the tool. A test execution tool does not replace the need for good test design and should not be used for every test. Some tests are still better executed manually. For example, a test that takes a very long time to automate and will not be run very often is better done manually.

Dependency on the tool vendor or community support

A commercial tool vendor may not be able to continue to support a tool that you have come to rely on. If a tool is acquired by a different organization, they may not have the same priorities. Sometimes tools that have been very popular are phased out. Open source tools also change over time and may become less suitable for you. When you choose a tool, always think about how you would cope if you had to change to a different tool, and structure your automation to cope with this, even if it seems very unlikely now.

Even an existing tool vendor or open source community may not provide good support forever. They may not respond quickly or adequately to questions, problems or bugs found in the tools (yes it does happen!). Upgrades may cause problems for you because of the way that you have used the tool, but the vendor may not be interested in helping you if you are in a small minority of their customers.

Using open source software

Although there is not a purchase price for open source tools, this doesn't mean they are cost-free. You still need to implement the tool and fit it into your own work practices. Since open source tools are very much supported by a community, if the people involved are no longer active, it may be difficult to get help with tool problems. If an open source product is abandoned, there may be no support for using it and no further updates. On the other hand, because many people may be involved, there may be a lot of updates to internal components of the tool, not all of which may be relevant to you.

Incompatible with development platform

Your own applications are changing over time as well, moving onto new platforms and using new technology. If you are in the forefront of your industry, your testing tools may not yet be able to deal with these innovations. Also, if developers move to a new development platform, this may cause problems for the existing automation and may even cause it to be abandoned.

Regulatory or safety requirements compliance

There is a risk that a tool will be suitable in other respects but may not comply with regulatory requirements and/or safety standards. Some regulatory bodies require that the tools themselves are approved. If the people who chose the tool were unaware of tool restrictions within their industry an unsuitable tool may be chosen.

6.3 EFFECTIVE USE OF TOOLS (ADDITIONAL MATERIAL)

In the previous Syllabus, there was a section about effective use of tools. Although this has now been removed from the current v4.0 Syllabus, we think that this section, taken from the previous book, contains advice and information that may still be useful to our readers. In this section, we discuss the principles and process of introducing tools into your organization. We look at the process of tool selection. We'll talk about how to carry out tool pilot projects. We'll conclude with thoughts on what makes tool introduction successful.

Lots of tool advice is online, including the Graham and Gamba wiki (2018) and good advice on automation can be found in Fewster and Graham [1999], Graham and Fewster [2012], Gamba and Graham [2018] and Axelrod [2018].

6.3.1 Main principles for tool selection

The place to start when introducing a tool into an organization is not with the tool: it is with the organization. In order for a tool to provide benefit, it must match a need within the organization and solve that need in a way that is both effective and efficient. The tool should help to build on the strengths of the organization and address its weaknesses. The organization needs to be ready for the changes that will come with the new tool. If the current testing practices are not good and the organization is not mature, then it is generally more cost-effective to improve testing practices rather than to try to find tools to support poor practices. Automating chaos just gives faster chaos!

Of course, we can sometimes improve our own processes in parallel with introducing a tool to support those practices. We can pick up some good ideas for improvement from the ways that the tools work. However, be aware that the tool should not take the lead but should provide support to what your organization defines.

The following factors are important in selecting a tool:

- Assessment of the organization's maturity (for example, strengths and weaknesses, readiness for change).
- Identification of the areas within the organization where tool support will help to improve testing processes.
- Understanding the technologies used by the test object(s), so that a tool will be selected that is compatible with those technologies.
- Knowledge of any build and continuous integration tools already being used within the organization, to make sure that the new tool(s) will integrate with them and be compatible.
- Evaluation of tools against clear requirements and objective criteria.
- Consideration of any free trial period for the tool (for commercial tools) to ensure that this gives adequate time to evaluate the tool.
- Evaluation of the vendor (including training, support and other commercial aspects) or support for non-commercial tools (open source).
- Identification of internal requirements for coaching and mentoring in the use of the tool.
- Evaluation of training needs for those who will use the tools directly and indirectly (for example, without technical detail), taking into account testing skills and test automation skills (for those working directly with the tools).
- Consideration of pros and cons of different licensing models (for example, commercial or open source).
- Estimation of a cost-benefit ratio based on a concrete and realistic business case (if required).

A final step would be to do a proof-of-concept evaluation. The purpose of this is to see whether or not the tool performs as expected, both with the software under test and with any other tools or aspects of your own infrastructure. For example, you want to find out now if your proposed tool cannot communicate with your configuration management system. You may also need to identify changes to your current infrastructure to enable the new tool(s) to be used effectively.

Tools versus people

There are some things that people can do much better or easier than a computer can. For example, when you see a friend in an unexpected place, say in an airport, you can immediately recognize their face. This is an example of pattern recognition that people are very good at, but it is not easy to write software that can recognize a face.

There are other things that computers can do much better or more quickly than people can. For example, can you add up 20 three-digit numbers quickly? This is not easy for most people to do, so you are likely to make some mistakes even if the numbers are written down. A computer does this accurately and very quickly. As another example, if people are asked to do exactly the same task over and over, they soon get bored and then start making mistakes.

The point is that it is a good idea to use computers to do things that computers are really good at and that people are not very good at. Tool support is very useful for repetitive tasks—the computer doesn't get bored and will be able to exactly repeat what was done before. Because the tool will be fast, this can make those activities much more efficient and more reliable. The tools can also do things that might overload a person, such as comparing the contents of a large data file or simulating how the system would behave.

6.3.2 Pilot project

After selecting a tool (or tools) and a successful proof-of-concept, the next step is to have a pilot project as the first step in using the tool(s) for real. The pilot project will use the tool in earnest but on a small scale, with sufficient time to explore different ways of using the tool. Objectives should be set for the pilot in order to assess whether or not the tool can accomplish what is needed within the current organizational context.

A pilot tool project should expect to encounter problems. They should be solved in ways that can be used by everyone later on. The pilot project should experiment with different ways of using the tool. For example, it should result in different reports from a test management tool, different scripting and comparison techniques for a test execution tool or different load profiles for a performance testing tool.

The objectives for a pilot project for a newly acquired tool are:

- To gain in-depth knowledge about the tool (more detail, more ways of using it) and to understand more fully both its strengths and its weaknesses.

- To see how the tool would fit with existing processes and practices, determining how those would need to adapt and change to work well with the tool, and how to use the tool to streamline and improve existing processes.

- To decide on standard ways of using the tool that will work for all potential users (for example, naming conventions for files and tests, creation of libraries, defining modularity, where different elements will be stored, how they and the tool itself will be maintained, and coding standards for test scripts).

- To assess whether or not the benefits will be achieved at reasonable cost.
- To understand (and experiment with) metrics that you want the tool(s) to collect and to report, and configuring the tool(s) to ensure that your goals for these metrics can be achieved.

6.3.3 Success factors for tools

Success is not guaranteed or automatic when acquiring a testing tool, but many organizations have succeeded. After a successful selection process and a pilot project, two other things are also important to get the greatest benefit from the tools: the way in which the tool(s) is deployed within the wider organization, and the way in which ongoing support is organized. Here are some of the factors that contribute to success:

- Rolling out the tool incrementally (after the pilot) to the rest of the organization (a gradual uptake of the tool, not trying to get the whole organization to use it at once immediately).
- Adapting and improving processes, testware and tool artefacts to get the best fit and balance between them and the use of the tool.
- Providing adequate support, training (for example, for those using the tool directly), coaching (for example, from external specialist automation consultants) and mentoring for tool users.
- Defining and communicating guidelines for the use of the tool, based on what was learned in the pilot (for example, internal standards for automation).
- Implementing a way to gather information about the use of the tool, to enable continuous improvement as tool use spreads through more of the organization.
- Monitoring the use of the tool and the benefits achieved, and adapting the use of the tool to take account of what is learned.
- Providing continuing support for anyone using test tools, such as the test team (for example, technical expertise is needed to help non-programmer testers who use keyword-driven testing).
- Gathering lessons learned, based on information gathered from all teams who are using test tools.

Any tools also need to be integrated both technically and organizationally into the software development life cycle. This may involve separate organizations that are responsible for different aspects, such as operations and/or third-party suppliers. Failure in tool use can come from any one factor; success needs all factors to be working together.

CHAPTER REVIEW

Let's review what you have learned in this chapter.

From Section 6.1, you should now be able to classify different types of test tools according to the test process activities that they support. The tool types you should now recognize are:

- management tools
- static testing tools
- test design and implementation tools
- test execution and coverage tools
- non-functional testing tools
- DevOps tools
- collaboration tools
- tools supporting scalability and deployment standardization
- other tools that assist in testing (e.g. spreadsheets).

From Section 6.2, you should be able to summarize the potential benefits and potential risks of test automation.

From this section, you should know the Glossary term **test automation**.

SAMPLE EXAM QUESTIONS

Question 1 Which tools help to support static testing?

a. Static analysis tools and test execution tools.

b. Review tools, static analysis tools and coverage measurement tools.

c. Performance testing tools and review tools.

d. Review tools and static analysis tools.

Question 2 A defect management tool is an example of which type of test tool?

a. DevOps tool.

b. Non-functional testing tool.

c. Management tool.

d. Collaboration tool.

Question 3 What are the potential benefits from using test automation?

a. Greater quality of code, reduction in the number of testers needed, better objectives for testing.

b. Greater repeatability of tests, reduction in repetitive work, objective assessment.

c. Greater responsiveness of users, reduction of tests run, objectives not necessary.

d. Greater quality of code, reduction in paperwork, fewer objections to the tests.

Question 4 What are potential risks of using test automation?

a. Over-reliance on tools, choosing an unsuitable tool.

b. Choosing the right tool vendor, reduced execution time.

c. Realistic expectations about the tool, developer involvement in scripting.

d. Providing measures too complex for humans, testers will spend more time on testing than in developing the automation.

CHAPTER SEVEN

ISTQB Foundation Exam

We wrote (and re-wrote and re-wrote) this book specifically to cover the International Software Testing Qualification Board (ISTQB) Foundation Syllabus v4.0 (2023). Because of this, mastery of the previous six chapters should ensure that you master the exam too. Here are some thoughts on how to make sure you show how much you know when taking the exam.

7.1 PREPARING FOR THE EXAM

7.1.1 Studying for the exam

Whether you have taken a preparatory course along with reading this book or just read the book, we have some study tips for certification candidates who intend to take an exam under one of the ISTQB-recognized Member Boards or Exam Providers.

Of course, you should carefully study the Syllabus. If you encounter any statements or concepts you do not *completely* understand, refer back to this book and to any course materials used, to prepare for the exam. Exam questions often turn on *precise* understanding of a concept or the wording.

While you should study the whole Syllabus, scrutinize the learning objectives in the Syllabus one by one. Ask yourself if you have achieved that learning objective to the given level of knowledge. If not, go back and review the appropriate material in the section corresponding to that learning objective. Note that the sections in both the Syllabus and the book often have the same numbering, and this is usually the same as the learning objective in the Syllabus.

Going beyond the Syllabus, notice that a number of international standards are referred to in the Syllabus. Some exam questions may be about the standards and there will also be questions about Glossary terms, and often even experienced testers are unfamiliar with them.

We recommend that you try to answer all the sample exam questions in this book. If you have understood the material in a chapter, you should have no trouble with the questions. Make sure you understand why the right answer is the right answer and why the wrong answers are wrong. If you cannot answer a question and don't understand why, review that section.

We also recommend that you try to do all the exercises in the book. If you can complete them correctly, you probably understand the concepts. If you do the exercises well, you are more likely to pass the exam.

Finally, be sure to take the mock exam we have provided at the end of this chapter. If you have understood the material in the book, you should have no trouble with most, if not all, of the questions on the mock exam. Make sure you carefully review the book sections corresponding to any questions that you haven't answered correctly. When you take the mock exam, try to simulate exam conditions as much as possible. Set aside a full hour to do the exam, and make sure that you are not disturbed during the hour: no interruptions! Turn off email and leave your phone in another room.

7.1.2 The exam and the Syllabus

The ISTQB Foundation exam is designed to assess your knowledge and understanding of basic testing ideas and terms, which provides a solid starting point for your future testing efforts. These exams are not perfect. Even well-qualified candidates often fail a few questions or disagree with some answers. Study hard, so you can take the exam in a relaxed and confident way.

If you take the exam unprepared, you should expect your lack of preparation to show in your score. Since you will pay to take the exam, consider preparation time an investment in protecting your exam fee.

The ISTQB Foundation Syllabus v4.0 (2023) is the current Syllabus. It replaces the 2018 version, which in turn replaced the 2011 version, which in turn replaced the 2007 version, which in turn replaced the 2005 version. Do not refer to the old Syllabi, as the exam covers the new Syllabus.

7.1.3 Where should you take the exam?

If you have taken an accredited training course, it is possible the exam will occur on the last day of the course. The exam fee may have been included in the course fee. If so, you should plan to study each evening during the course, including material that will be covered on the last day. Any topic covered in the Syllabus may be in the exam. You might even want to read this book before starting the course.

You might be taking an online course. Again, we recommend that you study the materials as you cover each section. Consider reading the book during or even before taking the course.

If you have studied on your own, without taking a course, then you will need to find a place open to the general public where the exam is offered. The exam is also available online from Pearson Vue and other exam administration companies. The website for the testing board for your country (the Member Board) should be able to help you find the right venue or company for a public or online exam.

In addition, ISTQB-recognized exam providers will provide you with ISTQB-branded certificates, as they work in close cooperation with the ISTQB and its Member Boards. The ISTQB and each ISTQB-recognized Member Board and Exam Provider are constitutionally obliged to support a single, universally accepted, international qualification scheme. Therefore, you are free to choose ISTQB-accredited trainers, if desired, and ISTQB-accredited exam providers based on factors like convenience, competence, suitability of the training to your immediate professional and business needs and price.

7.2   TAKING THE EXAM

All the Member Boards offer the same type of Foundation exam. It is a multiple choice exam that conforms to the guidelines established by the ISTQB. The exam in this book also follows the new guidelines from ISTQB for the balance of questions from the different sections/chapters. For some of us, it's been a while since we last took an exam and you may not remember all the tricks and techniques for taking a multiple choice exam. Here are a few paragraphs that will give you some ideas on how to take the exam.

7.2.1 How to approach multiple choice questions

Remember that there are two aspects to correctly answering a multiple choice exam question. First, you must understand the question and decide what the right answer is. Second, you must correctly communicate your answer by selecting the correct option from the choices listed.

Most questions have only one correct answer. This answer is always or most frequently True, given the stated context and conditions in the question. The other answers are intended to mislead people who do not completely understand the concept and are called 'distracters' in the examination business. However, beware there are some exam questions where two correct answers need to be chosen. So read the question carefully to find out what is required. For example, the ISTQB practice exams contain questions where you are asked to choose two options. You need to get both of them correct to get the point for that question.

Read the question carefully and make sure you understand it before you decide on your answer. It may help to highlight or underline the keywords or concepts stated in the question. Some questions may be ambiguous or ask which is the best or worst alternative. You should choose the best answer based on what you have learned in this book. Remember that your own situation may be different from the most common situation or the ideal situation.

Some of the distracters may confuse you. In other words, you might be unsure about the correct answers to some of the questions. So make two passes through the exam. On your first pass, answer all the questions you are sure of. Come back to the others later.

If a question will take a long time to answer, come back to it later, when you feel confident that you have enough time. Each correct answer is worth one point, whether it takes 30 seconds or 5 minutes to decide on it.

It's important to pace yourself. We suggest that you spend 45–60 seconds on each question. That way your first pass will take less than 40 minutes. You can then use the remaining time to have a one-minute break, then double-check your answers, and then finally go back to any questions you haven't answered yet.

If you use an answer sheet for the exam, they may vary slightly from one Member Board or Exam Provider to another. You should make sure that each answer corresponds to the correct question number. Especially if you have skipped a question, it is easy to get confused and mark the answer option above or below the question you are trying to answer.

Also, make sure you have an answer for every question. Double-check that you have selected the answer you want, as it is easy to select the wrong answer.

7.2.2 On trick questions

There are no deliberately designed trick questions in the ISTQB exams, but some can seem quite difficult if you are not completely sure of the right answer. Remember, just because it is hard to answer does not make it a trick. Here are some ideas for dealing with the tough questions.

First, read the question and each of the options very thoroughly. It is easy to read what you expect to be there rather than what is there.

If you are tempted to change your answer, do so only if you are quite sure. When you are undecided between two options, it's usually best to go with your first instinct. Remember that the correct answer is the one always or most often true. Simply because you can imagine a circumstance under which an answer might not be true does not make it wrong, just as knowing a 95-year-old cigarette smoker does not prove that cigarette smoking is not risky.

If you are taking an exam in person and using a paper copy, you are allowed to write on the question paper or make notes on paper (which would need to be handed in). In addition, there is typically no penalty for the wrong answer. So feel free to guess!

While the ISTQB guidelines call for straightforward questions, some topics are difficult to cover without some amount of complexity. Be especially careful with negative questions and complex questions. If the question asks which is False (or True), mark them F or T first if you can. This will make it easier to see which is the odd one out. If the question requires matching a list of definitions, you might be able to draw lines, when doing a paper-based exam, between the definitions and the words they define. And remember to use the process of elimination to work for you whenever possible. Use what you know to eliminate wrong answers and cross them out up front to avoid mistakes.

Finally, remember that the ISTQB exam is about the theory of good practice, not what you typically do in practice. The ISTQB Syllabus was developed by an international working group of testing experts based on the good practices they have seen in the real world. In testing, good practices may lead typical practices by about 20 years' time. Some organizations are struggling to implement even basic testing principles, such as removing the misconception that 'complete' testing is possible, while other organizations may already be doing most of the things discussed in the Syllabus. Still other organizations may do things well but differently from what is described in the Syllabus. So remember to apply what you have learned in this book to the exam. When an exam question calls into conflict what you have learned here and what you typically have done in the past, rely on what you have learned to answer the exam question. And then go back to your workplace and put the good ideas that you have learned into practice!

7.2.3 Last but not least

ISTQB has sample exam papers available to download, as well as the correct answers and justifications for those answers. On the main ISTQB website, scroll down and choose 'Foundation Level, Certified Tester'. The Syllabus and practice exams available to download should be shown on the page.

We wish you good luck with your certification exam and best of success in your career as a test professional! We stand poised on the brink of great forward progress in the field of testing and are happy that you will be a part of it.

7.3 MOCK EXAM

In the real exam, you will have 60 minutes (75 if the exam is not in your first language) to work through 40 questions of approximately the same difficulty mix and Syllabus distribution as shown in the following mock exam. After you have taken this mock exam, check your answers with the answer key. The answers to all exam questions in this book are in the section after the Glossary. Note that this mock exam is fully compliant with ISTQB exam structure and rules.

Fundamentals of Testing

Question 1 Consider the following statements about testing and debugging and identify which are True:

1. Debugging is a testing activity.

2. Testers may be involved in debugging and component testing in Agile development.

3. Testing finds, analyzes and fixes defects.

4. Debugging executes tests to show failures caused by defects.

5. Checking whether fixes resolved the defects found is a form of testing.

a. 2 and 5.

b. 2, 3 and 4.

c. 1, 2 and 5.

d. Only 5 is True.

Question 2 Consider the following factors:

1. The test levels and test types being considered.

2. The best practices implemented in other companies.

3. The business domain.

4. Required internal and external standards.

5. The way in which the previous system was developed.

Which of these factors may influence the test process in context?

a. 1, 2 and 5.

b. 2, 4 and 5.

c. 1, 3 and 4.

d. 1, 3 and 5.

Question 3 Consider the following list of test activities:

I.   Analysis and design.

II. Test completion activities.

III. Implementation.

IV. Planning.

V.   Execution.

Which of the following places these in their logical sequence?

a. I, II, III, IV and V.

b. IV, I, V, III and II.

c. IV, I, III, V and II.

d. I, IV, V, III and II.

Question 4 System test execution on a project is planned for eight weeks. After a week of testing, a tester suggests that the test objective stated in the test plan of 'finding as many defects as possible during system test' might be more closely met by redirecting the test effort in what way?

a. By asking a selection of users what is most important for them, and testing that.

b. By testing the main workflows of the business.

c. By repeating the unit and integration tests.

d. By testing in areas where the most defects have already been found.

Question 5 Which of the following statements regarding the whole-team approach is NOT True?

a. The whole team approach improves team dynamics.

b. The whole team approach enhances communication and collaboration within the team.

c. The whole team approach creates synergy by allowing the various skill sets within the team to be leveraged for the benefit of the project.

d. The whole team approach ensures a high level of test independence.

Question 6 Which of the following is an advantage of independent testing?

a. Independent testers do NOT have to spend time communicating with the project team.

b. Developers can stop worrying about the quality of their work and focus on producing more code.

c. The others on a project can pressure the independent testers to accelerate testing at the end of the schedule.

d. Independent testers sometimes question the assumptions behind requirements, designs and implementations.

Question 7 A regression set is run on a weekly basis to mitigate the risk of changes having an adverse effect on this system. The regression risk is managed successfully with this approach; however, the number of defects found running the regression test set has decreased over time.

Which testing principle is implicitly referred to in this description?

a. Testing shows the presence, not the absence of defects.

b. Early testing saves time and money.

c. Tests wear out.

d. Absence-of-defects fallacy.

Question 8 Which of the tasks below is typically done by a tester (not a test manager)?

a. Support setting up the defect management system.

b. Share testing perspectives with other project activities such as integration planning.

c. Design, set up and verify test environment(s).

d. Initiate the analysis, design, implementation and execution of tests.

Testing throughout the Software Development Lifecyle

Question 9 Consider the following statements about regression tests:

I.   They may usefully be automated if they are well designed.

II. They are the same as confirmation tests (re-tests).

III. They are a way to reduce the risk of a change having an adverse effect elsewhere in the system.

IV. They are only effective if automated.

Which pair of statements is True?

a. I and II.

b. I and III.

c. II and III.

d. II and IV.

Question 10 Which statement about functional, non-functional and white-box testing is True?

a. Functional testing evaluates characteristics such as reliability, security or usability; non-functional testing evaluates characteristics such as system architecture and the thoroughness of testing; white-box testing evaluates characteristics such as completeness and correctness.

b. Functional testing evaluates characteristics such as completeness and correctness; non-functional testing evaluates characteristics such as reliability, security or usability; white-box testing evaluates characteristics such as system architecture and the thoroughness of testing.

c. Functional testing evaluates characteristics such as completeness and correctness; non-functional testing evaluates characteristics such as system architecture and the thoroughness of testing; white-box testing evaluates characteristics such as reliability, security or usability.

d. Functional testing evaluates characteristics such as system architecture and the thoroughness of testing; non-functional testing evaluates characteristics such as reliability, security or usability; white-box testing evaluates characteristics such as completeness and correctness.

Question 11 Which of the following statements is an example of a good testing practice that applies to all software development lifecycles?

a. For every software development activity, there is a corresponding test activity.

b. Tests are written first; the code is written to satisfy the tests.

c. The risk in regression is minimized due to a large set of automated regression tests.

d. Non-functional security testing is starting to be performed at component level.

Question 12 With which test-first approach to development are tests derived from acceptance criteria as part of the system design process?

a. Test-Driven Development (TDD).

b. Acceptance Test-Driven Development (ATDD).

c. Behaviour Driven-Development (BDD).

d. Both Acceptance Test-Driven Development (ATDD) and Behaviour Driven-Development (BDD).

Question 13 What is the key aspect of the shift-left approach to testing?

a. Experience-based testing should be performed at the component testing level.

b. Test automation should be performed using a white-box coverage technique.

c. Testing is performed earlier in the lifecycle.

d. Requirements should always be reviewed using a formal review technique.

Question 14 Which of the following statements about retrospectives is True regarding how they can be used as a mechanism for improvement?

a. During retrospectives, the tester should avoid commenting on non-testing activities and focus their contribution only on testing activities.

b. Potential changes to the way-of-working aimed at improving the quality of the test basis can be discussed during retrospectives.

c. Retrospectives are held in the middle of each iteration in order to highlight and discuss impediments that are blocking progress.

d. Unlike formal reviews, retrospectives do not take advantage of conducting follow-up activities.

■■■■■■■■■■■■■■■■■■■■■■■■■■■■

Static Testing

Question 15 Which of the following statements about static testing is False?

a. Reviews are applicable to web pages and user guides.

b. Static analysis is applicable to executing test scripts.

c. Static analysis can be applied to work products written in natural language (e.g. English).

d. Reviews can be applied to security requirements and project budgets.

Question 16 Which of the following statements are characteristics of static testing, and which are characteristics of dynamic testing?

1. Finds defects in work products directly.

2. Better for ensuring internal quality (e.g. standards are followed).

3. Finds defects through failures in execution.

4. Easier and cheaper to find and fix security vulnerabilities (e.g. buffer overflow).

5. Focuses on externally visible behaviours.

a. Static testing: 2 and 4, Dynamic testing: 1, 3 and 5.

b. Static testing: 3 and 5, Dynamic testing: 1, 2 and 4.

c. Static testing: 1, 2 and 4, Dynamic testing: 3 and 5.

d. Static testing: 1, 2 and 3, Dynamic testing: 4 and 5.

Question 17 What is the most important factor for successful performance of reviews?

a. A separate scribe during the logging meeting.

b. Trained participants and review leaders.

c. The availability of tools to support the review process.

d. A reviewed test plan.

Question 18 What factor should you NOT consider when choosing the type of review to use for a particular work product?

a. Any legal or regulatory requirements.

b. The work product has not previously been reviewed.

c. The people available to perform the review.

d. The purpose of reviewing the work product at this time.

■■■■■■■■■■■■■■■■■■■■■■■■■■■■

Test Analysis and Design

Question 19 Assume postal rates for light letters are:

$0.25 up to 10 grams.

$0.35 up to 50 grams.

$0.45 up to 75 grams.

$0.55 up to 100 grams.

Which test inputs (in grams) could be selected using equivalence partitioning?

a. 0, 9, 10, 49, 50, 74, 75, 99, 100.

b. 5, 35, 65, 95, 115.

c. 0, 1, 10, 11, 50, 51, 75, 76, 100, 101.

d. 5, 25, 35, 45, 55.

Question 20 How do experience-based test techniques differ from black-box test techniques?

a. They depend on the tester's understanding of the way the system is structured rather than on a documented record of what the system should do.

b. They depend on an individual's domain knowledge and expertise rather than on a documented record of what the system should do.

c. They depend on a documented record of what the system should do rather than on an individual's personal view.

d. They depend on having older testers rather than younger testers.

Question 21 If you are flying with an economy ticket, there is a possibility that you may get upgraded to business class, especially if you hold a gold card in the airline's frequent flyer programme. If you do not hold a gold card, there is a possibility that you will get bumped off the flight if it is full and you check in late. This is shown in Figure 7.1. Note that each box (i.e. statement) has been numbered.

Three tests have already been run:

Test 1: Gold card holder who gets upgraded to business class.

Test 2: Non-gold card holder who stays in economy.

Test 3: A person who is bumped from the flight.

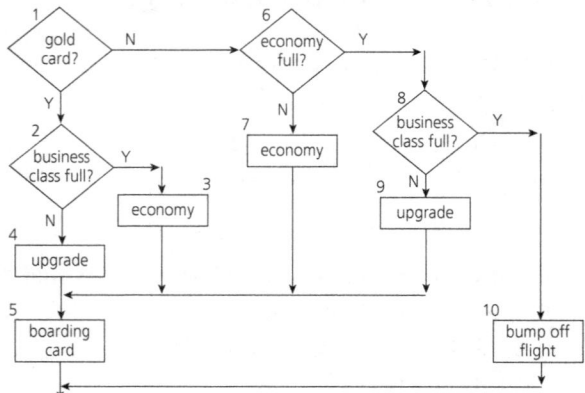

FIGURE 7.1 Control flow diagram for flight check-in

What additional tests would be needed to achieve 100% branch coverage?

a. A gold card holder who stays in economy and a non-gold card holder who gets upgraded to business class.

b. A gold card holder and a non-gold card holder who are both upgraded to business class.

c. A gold card holder and a non-gold card holder who both stay in economy class.

d. A gold card holder who is upgraded to business class and a non-gold card holder who stays in economy class.

Question 22 Which of the following statements about error guessing are True?

P. Error guessing uses a test charter and test session sheets.

Q. Error guessing is based on tester knowledge.

R. Error guessing always uses a checklist.

S. Error guessing is a black-box test technique.

T. Error guessing can be based on mistakes developers may make.

a. P and T.

b. Q and R.

c. Q and T.

d. Q and S.

Question 23 If the temperature falls below 18 degrees, the heating is switched on. When the temperature reaches 21 degrees, the heating is switched off. What is the minimum set of test input values to cover all valid equivalence partitions?

a. 15, 19 and 25 degrees.

b. 17, 18, 20 and 21 degrees.

c. 18, 20 and 22 degrees.

d. 16 and 26 degrees.

Question 24 Assume postal rates for light letters are:

$0.25 up to 10 grams.

$0.35 up to 50 grams.

$0.45 up to 75 grams.

$0.55 up to 100 grams.

Which test inputs (in grams) would be selected using boundary value analysis?

a. 0, 9, 19, 49, 50, 74, 75, 99, 100.

b. 10, 50, 75, 100, 250, 1000.

c. 0, 1, 10, 11, 50, 51, 75, 76, 100, 101.

d. 25, 26, 35, 36, 45, 46, 55, 56.

Question 25 Consider Table 7.1.

TABLE 7.1 Decision table for car rental

Conditions	Rule 1	Rule 2	Rule 3	Rule 4
Over 23?	F	T	T	T
Clean driving record?	Don't care	F	T	T
On business?	Don't care	Don't care	F	T
Actions				
Supply rental car?	F	F	T	T
Premium charge?	F	F	F	T

Given this decision table, what is the expected result for the following test cases?

TC1: A 26-year-old on business but with violations or accidents on their driving record.

TC2: A 62-year-old tourist with a clean driving record.

a. TC1: Do not supply car;
TC2: Supply car with premium charge.

b. TC1: Supply car with premium charge;
TC2: Supply car with no premium charge.

c. TC1: Do not supply car;
TC2: Supply car with no premium charge.

d. TC1: Supply car with premium charge;
TC2: Do not supply car.

Question 26 What is exploratory testing?

a. The process of anticipating or guessing where defects might occur.

b. A systematic approach to identifying specific equivalent classes of input.

c. The testing carried out by a chartered engineer.

d. Concurrent test design, test execution, test logging and learning.

Question 27 Consider the state transition diagram in Figure 7.2.

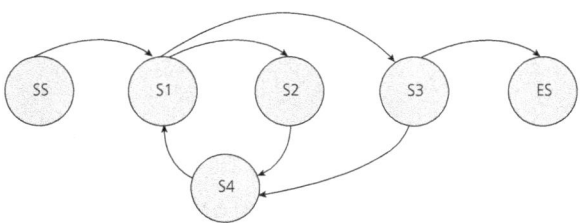

FIGURE 7.2 State transition diagram

Given this diagram, which test case below covers every valid transition?

a. SS – S1 – S2 – S4 – S1 – S3 – ES

b. SS – S1 – S2 – S3 – S4 – S3 – S4 – ES

c. SS – S1 – S2 – S4 – S1 – S3 – S4 – S1 – S3 – ES

d. SS – S1 – S4 – S2 – S1 – S3 – ES

Question 28 What statement about collaborative authorship of user stories is True?

a. Decision tables and state transition techniques should be used.

b. Brainstorming and mind-mapping techniques can be used.

c. A shared vision is defined by the business stakeholder(s).

d. If the user story is not clear, it is up to the testers to ensure that it is testable.

Question 29 Which of the following statements are True for the value of white-box testing?

V. White-box testing can detect omissions from user stories or requirements.

W. White-box testing is based on the software implementation.

X. White-box testing can use requirements or user stories to determine test inputs.

Y. White-box testing can use requirements or user stories to determine expected outputs.

Z. White-box testing is related to confidence in how well the system has been tested.

a. V, X and Z

b. W, Y and Z

c. V, W and Y

d. X, Y and Z

Managing the Test Activities

Question 30 Consider the following list of either product or project risks:

I.   An incorrect calculation of fees might short-change the organization.

II. A vendor might fail to deliver a system component on time.

III. A defect might allow hackers to gain administrative privileges.

IV. A skills gap might occur in a new technology used in the system.

V.   A defect-prioritization process might overload the development team.

Which of the following statements is True?

a. I is primarily a product risk and II, III, IV and V are primarily project risks.

b. II and V are primarily product risks and I, III and V are primarily project risks.

c. I and III are primarily product risks, while II, IV and V are primarily project risks.

d. III and V are primarily product risks, while I, II and IV are primarily project risks.

Question 31 Review the following portion of a defect report, which occurs after the defect ID, time and date, and author.

1. I place any item in the shopping cart.

2. I place any other (different) item in the shopping cart.

3. I remove the first item from the shopping cart but leave the second item in the cart.

4. I click the <Checkout> button.

5. I expect the system to display the first checkout screen. Instead, it gives the pop-up error message, 'No items in shopping cart. Click <Okay> to continue shopping'.

6. I click <Okay>.

7. I expect the system to return to the main window to allow me to continue adding and removing items from the cart. Instead, the browser terminates.

8. The failure described in steps 5 and 7 occurred in each of three attempts to perform steps 1, 2, 3, 4 and 6.

Assume that no other narrative information is included in the report. Which of the following important aspects of a good defect report is missing from this report?

a. The steps to reproduce the failure.

b. The summary.

c. The check for intermittence.

d. The use of an objective tone.

Question 32 Which of the following is a test metric?

a. Percentage of planned functionality that has completed development.

b. Confirmation test results (pass/fail).

c. Cost of development, including time and effort.

d. Effort spent in fixing defects, not including confirmation testing or regression testing.

Question 33 A test plan is written specifically to describe a level of testing where the primary goal is establishing confidence in the system. Which of the following is a likely name for this document?

a. Master test plan.

b. System test plan.

c. Acceptance test plan.

d. Project plan.

Question 34 Consider the following activities that might relate to configuration management:

I.   Identify and document the characteristics of a test item.

II. Control changes to the characteristics of a test item.

III. Check a test item for defects introduced by a change.

IV. Record and report the status of changes to test items.

V.   Confirm that changes to a test item fixed a defect.

Which of the following statements is True?

a. Only I is a configuration management task.

b. All are configuration management tasks.

c. I, II and III are configuration management tasks.

d. I, II and IV are configuration management tasks.

Question 35
You are the test manager for a user acceptance test project. You have decided to use a requirements-based and risk-based test strategy. Test suites have been designed to cover all the requirements.

Table 7.2 shows the priority and the designed test suites associated with the requirements (a lower priority number means higher risk).

The table also shows the configuration of the test environment in which the requirements can be tested. Thus, two different configurations, ABC and XYZ, are required to execute all test suites.

Assume that the test environment is delivered initially to the test team under the configuration ABC; XYZ will be available from the start of week 4. Each test suite will take approximately 1 week to execute.

The execution of TS4 depends on TS3 to be executed before, and the execution of TS2 depends on TS1 to be executed before.

Based only on the given information, which of the following test execution schedules would be preferred given the constraints of the test environment and the test strategy chosen?

a. TS5, TS6, TS4, TS3, TS1, TS2, TS7.

b. TS5, TS4, TS1, TS6, TS3, TS2, TS7.

c. TS5, TS6, TS1, TS3, TS4, TS2, TS7.

d. TS5, TS6, TS1, TS2, TS7, TS4, TS3.

TABLE 7.2 Priority and dependency table for Question 35

Requirement	Priority	Test suites	Environment	Dependency
R1	4	TS1	ABC	
R2	4	TS2	ABC	TS1
R3	3	TS3	XYZ	
R4	2	TS4	XYZ	TS3
R5	1	TS5, TS6	ABC	
R6	5	TS7	ABC	

Question 36 Which of the following statements is True about the testing quadrants?

a. Quadrant one tests should be manual.

b. Quadrant two tests can use documented test oracles.

c. Quadrant three tests include all non-functional tests.

d. Quadrant four tests include all non-functional tests.

Question 37 Which of the following is the best example of verbal communication of test status to fellow team members?

a. An automated dashboard generated by a test tool.

b. A report following the ISO 29119-3 standard.

c. A presentation of test results to executive management.

d. A tester reporting results in a daily stand-up meeting.

Question 38 Use estimation techniques to calculate the required test effort. In three past projects—releases 1, 2 and 3—the development effort was 150, 250 and 100 person-days, respectively. The test effort was 50, 80 and 40 person-days, respectively, on releases 1, 2 and 3. Release 4 is predicted to be 200 person-days. Which of the following correctly applies ratio estimation to predict the test effort for release 4?

a. The test effort is estimated at 57 person-days.

b. The test effort is estimated at 67 person-days.

c. The test effort is estimated at 80 person-days.

d. The test effort cannot be estimated via ratio estimation.

Test Tools

Question 39 Which type of test tool facilitates the generation of test cases and test data?

a. DevOps tool.

b. Test design and implementation tool.

c. Test execution and coverage tool.

d. Collaboration tool.

Question 40 Consider the following statements about test automation.

1. Tools can provide objective measurement of software.

2. Manual testing should be eliminated through the use of the most appropriate tools.

3. Once the test automation is in place, there is no need for continued investment of time.

4. Tools run tests more consistently and repeatably than human testers.

5. Using open source tools means that support will always be available.

Which of the above statements are True and which are False?

a. 1) and 4) are True; 2), 3) and 5) are False.

b. 2), 3) and 5) are True: 1) and 4) are False.

c. 1), 4) and 5) are True; 2) and 3) are False.

d. 1) and 5) are True; 2), 3) and 4) are False.

GLOSSARY

This Glossary provides the definition set of software testing terms used in the Foundation syllabus, as defined by ISTQB.

The Glossary has been arranged in a single section of definitions ordered alphabetically. Some terms are preferred to other synonymous ones, in which case the definition of the preferred term appears, with the synonyms listed afterwards. For example, a synonym of *white-box testing* is *structural testing.*

'Refer also to' cross-references are also used. They assist the user to quickly navigate to the right index term. 'Refer also to' cross-references are constructed for relationships such as broader term to a narrower term and overlapping meanings between two terms.

Finally, note that the terms that are underlined are those that are specifically mentioned as keywords in the Syllabus at the beginning of Chapters 1 to 6. These are the terms that you should know for the exam.

Acceptance criteria　The criteria that a component or system must satisfy in order to be accepted by a user, customer or other authorized entity.

Acceptance test-driven development　A collaboration-based test-first approach that defines acceptance tests in the stakeholders' domain language.

Acceptance testing　A test level that focuses on determining whether to accept the system. Refer also to: *User acceptance testing.*

Actual result　The behaviour produced/observed when a component or system is tested.
Synonym: actual outcome

Agile software development　A group of software development methodologies based on iterative incremental development, where requirements and solutions evolve through collaboration between self-organizing cross-functional teams. Refer also to: *Feature-driven development.*

Alpha testing　A type of acceptance testing performed in the developer's test environment by roles outside the development organization.

Anomaly　A condition that deviates from expectation.

API testing　Testing performed by submitting requests to the test object using its application programming interface.

Audit　An independent examination of a work product or process performed by a third party to assess whether it complies with specifications, standards, contractual agreements or other criteria.

Availability　The degree to which a component or system is operational and accessible when required for use.

Behaviour-driven development　A collaborative approach to development in which the team is focusing on delivering expected behaviour of a component or system for the customer, which forms the basis for testing.

Beta testing　A type of acceptance testing performed at an external site to the developer's test environment by roles outside the development organization.

Black-box test technique　A test technique based on an analysis of the specification of a component or system.
Synonyms: black-box test design technique, specification-based test technique

Black-box testing　Testing based on an analysis of the specification of the component or system.
Synonym: specification-based testing

Boundary value　A minimum or maximum value of an ordered equivalence partition. Refer also to: *Boundary value analysis.*

Boundary value analysis　A black-box test technique in which test cases are designed based on boundary values. Refer also to: *Boundary value.*

Branch　A transfer of control between two nodes in the control flow graph of a test item.

Branch coverage　The coverage of branches in a control flow graph.

Branch testing　A white-box test technique in which the test conditions are branches.

Cause-effect diagram　A graphical representation used to organize and display the interrelationships of various possible root causes of a problem.

Possible causes of a real or potential defect or failure are organized in categories and subcategories in a horizontal tree-structure, with the (potential) defect or failure as the root node. Synonyms: fishbone diagram, ishikawa diagram

Checklist-based review A review technique guided by a list of questions or required attributes.

Checklist-based testing An experience-based test technique in which test cases are designed to exercise the items of a checklist.

Coding standard A standard that describes the characteristics of a design or a design description of data or program components.

Collaboration-based test approach An approach to testing that focuses on defect avoidance by collaborating among stakeholders.

Compatibility The degree to which a component or system can exchange information with other components or systems and/or perform its required functions while sharing the same hardware or software environment.

Complexity The degree to which the design or code of a component or system is difficult to understand.

Compliance Adherence of a work product to standards, conventions or regulations in laws and similar prescriptions.

Component A part of a system that can be tested in isolation. Synonyms: module, unit

Component integration testing The integration testing of components. Synonyms: module integration testing, unit integration testing

Component testing A test level that focuses on individual hardware or software components. Synonyms: module testing, unit testing

Configuration management A discipline applying technical and administrative direction and surveillance to identify and document the functional and physical characteristics of a configuration item, control changes to those characteristics, record and report change processing and implementation status, and verify that it complies with specified requirements.

Confirmation testing A type of change-related testing performed after fixing a defect to confirm that a failure caused by that defect does not reoccur. Synonym: re-testing

Continuous integration An automated software development procedure that merges, integrates and tests all changes as soon as they are committed.

Continuous testing An approach that involves a process of testing early, testing often, test everywhere and automate to obtain feedback on the business risks associated with a software release candidate as rapidly as possible.

Control flow The sequence in which operations are performed by a business process, component or system.

Cost of quality The total costs incurred on quality activities and issues and often split into prevention costs, appraisal costs, internal failure costs and external failure costs.

Coverage The degree to which specified coverage items are exercised by a test suite, expressed as a percentage. Synonym: test coverage

Coverage criteria The criteria to define the coverage items required to reach a test objective. Refer also to: *Coverage item*.

Coverage item An attribute or combination of attributes derived from one or more test conditions by using a test technique. Refer also to: *Coverage criteria*.

Dashboard A representation of dynamic measurements of operational performance for some organization or activity, using metrics represented via metaphors such as visual dials, counters and other devices resembling those on the dashboard of an automobile, so that the effects of events or activities can be easily understood and related to operational goals.

Debugging The process of finding, analyzing and removing the causes of failures in a component or system.

Decision table testing A black-box test technique in which test cases are designed to exercise the combinations of conditions and the resulting actions shown in a decision table.

Defect An imperfection or deficiency in a work product where it does not meet its requirements or specifications. Synonyms: bug, fault

Defect density The number of defects per unit size of a work product. Synonym: fault density

Defect detection percentage   The number of defects found by a test level, divided by the number found by that test level and any other means afterwards.
Synonym: fault detection percentage

Defect management   The process of recognizing, recording, classifying, investigating, resolving and disposing of defects.

Defect report   Documentation of the occurrence, nature, and status of a defect.
Synonym: bug report

Definition   A description of the meaning of a word.

Driver   A component or tool that temporarily replaces another component and controls or calls a test item in isolation.
Synonym: test driver

Dynamic testing   Testing that involves the execution of the test item. Refer also to: *Static testing*.

Effectiveness   The extent to which correct and complete goals are achieved. Refer also to: *Efficiency*.

Efficiency   The degree to which resources are expended in relation to results achieved. Refer also to: *Effectiveness, Performance efficiency*.

Entry criteria   The set of conditions for officially starting a defined task. Refer also to: *Exit criteria*.

Equivalence partition   A subset of the value domain of a variable within a component or system in which all values are expected to be treated the same based on the specification.
Synonym: equivalence class

Equivalence partitioning   A black-box test technique in which test conditions are equivalence partitions exercised by one representative member of each partition.
Synonym: partition testing

Error   A human action that produces an incorrect result.
Synonym: mistake

Error guessing   A test technique in which tests are derived on the basis of the tester's knowledge of past failures, or general knowledge of failure modes.

Exhaustive testing   A test approach in which the test suite comprises all combinations of input values and preconditions.
Synonym: complete testing

Exit criteria   The set of conditions for officially completing a defined task. Refer also to: *Entry criteria*.

Synonyms: test completion criteria, completion criteria

Expected result   The observable predicted behaviour of a test item under specified conditions based on its test basis.
Synonyms: predicted outcome, expected outcome

Experience-based test technique   A test technique based on the tester's experience, knowledge and intuition.
Synonyms: experience-based test design technique, experience-based technique

Exploratory testing   An approach to testing in which the testers dynamically design and execute tests based on their knowledge, exploration of the test item and the results of previous tests. Refer also to: *Test charter*.

Failed   The status of a test result in which the actual result does not match the expected result.

Failure   An event in which a component or system does not perform a required function within specified limits.

Feature-driven development   An iterative and incremental software development process driven from a client-valued functionality (feature) perspective. Feature-driven development is mostly used in Agile software development. Refer also to: *Agile software development*.

Finding   A result of an evaluation that identifies some important issue, problem or opportunity.

Formal review   A type of review that follows a defined process with a formally documented output.

Functional appropriateness   The degree to which the functions facilitate the accomplishment of specified tasks and objectives.
Synonym: suitability

Functional completeness   The degree to which the set of functions covers all the specified tasks and user objectives.

Functional correctness   The degree to which a component or system provides the correct results with the needed degree of precision.
Synonym: accuracy

Functional testing   Testing performed to evaluate if a component or system satisfies functional requirements.

Heuristic   A generally recognized rule of thumb that helps to achieve a goal.

Impact analysis   The identification of all work products affected by a change, including an estimate of the resources needed to accomplish the change.

Incremental development model   A type of software development lifecycle model in which the component or system is developed through a series of increments.

Independence of testing   Separation of responsibilities, which encourages the accomplishment of objective testing.

Informal review   A type of review that does not follow a defined process and has no formally documented output.

Inspection   A type of formal review that uses defined team roles and measurement to identify defects in a work product, and improve the review process and the software development process.

Integration testing   A test level that focuses on interactions between components or systems.

Integrity The degree to which a component or system allows only authorized access and modification to a component, a system or data.

Iterative development model   A type of software development lifecycle model in which the component or system is developed through a series of repeated cycles.

Maintainability   The degree to which a component or system can be modified by the intended maintainers. Refer also to: *Testability.*

Maintenance   The process of modifying a component or system after delivery to correct defects, improve quality characteristics or adapt to a changed environment.

Maintenance testing   Testing the changes to an operational system or the impact of a changed environment to an operational system.

Maturity   (1) The capability of an organization with respect to the effectiveness and efficiency of its processes and work practices. (2) The degree to which a component or system meets needs for reliability under normal operation.

Mean time to failure   The average time from the start of operation to a failure for a component or system.

Measurement   The process of assigning a number or category to an entity to describe an attribute of that entity.

Metric   A measurement scale and the method used for measurement.

Moderator   (1) The person responsible for running review meetings. (2) The person who performs a usability test session.
Synonym: facilitator

N-switch coverage   The coverage of sequences of N+1 transitions.

Negative testing   Testing a component or system in a way for which it was not intended to be used.
Synonyms: invalid testing, dirty testing

Neuron coverage   The coverage of activated neurons in the neural network for a set of tests.

Non-functional testing   Testing performed to evaluate that a component or system complies with non-functional requirements.

Operational acceptance testing   A type of acceptance testing performed to determine if operations and/or systems administration staff can accept a system.
Synonym: production acceptance testing

Pair testing   A test approach in which two team members simultaneously collaborate on testing a work product.

Passed   The status of a test result in which the actual result matches the expected result.

Path   A sequence of consecutive edges in a directed graph.
Synonym: control flow path

Performance efficiency   The degree to which a component or system uses time, resources and capacity when accomplishing its designated functions. Refer also to: *Efficiency.*

Planning poker   A consensus-based estimation technique, mostly used to estimate effort or relative size of user stories in Agile software development. It is a variation of the Wideband Delphi method using a deck of cards with values representing the units in which the team estimates.

Portability   The degree to which a component or system can be transferred from one hardware, software or other operational or usage environment to another.

Postcondition   The expected state of a test item and its environment at the end of test case execution.

Precondition   The required state of a test item and its environment prior to test case execution.

Priority   The level of (business) importance assigned to an item, e.g., defect.

Product risk   A risk that impacts the quality of a product. Refer also to: *Risk*.

Project risk   A risk that impacts project success. Refer also to: *Risk*.

Quality   The degree to which a work product satisfies stated and implied needs of its stakeholders.

Quality assurance   Activities focused on providing confidence that quality requirements will be fulfilled.

Quality characteristic   A category of quality attributes that bears on work product quality. Synonyms: quality attribute, software quality characteristic, software product characteristic

Quality control   Activities designed to evaluate the quality of a component or system. Refer also to: *Testing*.

Quality risk   A product risk related to a quality characteristic.

Regression testing   A type of change-related testing to detect whether defects have been introduced or uncovered in unchanged areas of the software.

Regulatory acceptance testing   A type of acceptance testing performed to determine the compliance of the test object.

Reliability   The degree to which a component or system performs specified functions under specified conditions for a specified period of time.

Requirement   A provision that contains criteria to be fulfilled.

Retrospective   A regular event in which team members discuss results, review their practices and identify ways to improve. Synonyms: project retrospective, retrospective meeting, post-project meeting

Review   A type of static testing in which a work product or process is evaluated by one or more individuals to detect defects or to provide improvements.

Reviewer   A participant in a review who identifies defects in the work product. Synonyms: checker, inspector

Risk   A factor that could result in future negative consequences. Refer also to: *Product risk, Project risk*.

Risk analysis   The overall process of risk identification and risk assessment.

Risk assessment   The process to examine identified risks and determine the risk level.

Risk control   The overall process of risk mitigation and risk monitoring.

Risk identification   The process of finding, recognizing and describing risks.

Risk impact   The damage that will be caused if the risk becomes an actual outcome or event. Synonym: impact

Risk level   The measure of a risk defined by risk impact and risk likelihood. Synonym: risk exposure

Risk likelihood   The probability that a risk will become an actual outcome or event. Synonym: likelihood

Risk management   The process for handling risks.

Risk mitigation   The process through which decisions are reached and protective measures are implemented for reducing or maintaining risks to specified levels.

Risk monitoring   The activity that checks and reports the status of known risks to stakeholders.

Risk-based testing   Testing in which the management, selection, prioritization and use of testing activities and resources are based on corresponding risk types and risk levels.

Root cause   A source of a defect such that if it is removed, the occurrence of the defect type is decreased or removed.

Root cause analysis   An analysis technique aimed at identifying the root causes of defects. Synonym: causal analysis

Scalability   The degree to which a component or system can be adjusted for changing capacity.

Scenario-based review   A review technique in which a work product is evaluated to determine its ability to address specific scenarios.

Scribe   A person who records information at a review meeting. Synonym: recorder

Security The degree to which a component or system protects its data and resources against unauthorized access or use and secures unobstructed access and use for its legitimate users. Synonym: information security

Sequential development model A type of software development lifecycle model in which a complete system is developed in a linear way of several discrete and successive phases with no overlap between them.

Service virtualization A technique to enable virtual delivery of services which are deployed, accessed and managed remotely.

Session-based testing A test approach in which test activities are planned as test sessions.

Severity The degree of impact that a defect has on the development or operation of a component or system.

Shift-left An approach to performing testing and quality assurance activities as early as possible in the software development lifecycle.

Simulator A component or system used during testing which behaves or operates like a given component or system.

Smoke test A test suite that covers the main functionality of a component or system to determine whether it works properly before planned testing begins. Synonyms: sanity test, intake test, confidence test

Software development lifecycle The activities performed at each stage in software development and how they relate to one another logically and chronologically. Synonym: lifecycle model

State transition testing A black-box test technique in which test cases are designed to exercise elements of a state transition model. Synonym: finite state testing

Statement coverage The coverage of executable statements.

Statement testing A white-box test technique in which test cases are designed to execute statements.

Static analysis The process of evaluating a component or system without executing it, based on its form, structure, content or documentation.

Static testing Testing that does not involve the execution of a test item. Refer also to: *Dynamic testing*

Stub A skeletal or special-purpose implementation of a software component, used to develop or test a component that calls or is otherwise dependent on it. It replaces a called component.

System integration testing The integration testing of systems.

System testing A test level that focuses on verifying that a system as a whole meets specified requirements.

System under test A type of test object that is a system.

Technical review A formal review by technical experts that examine the quality of a work product and identify discrepancies from specifications and standards. Refer also to: peer review

Test A set of one or more test cases.

Test analysis The activity that identifies test conditions by analyzing the test basis.

Test approach The manner of implementing testing tasks.

Test automation The use of software to perform or support test activities.

Test automation framework A tool that provides an environment for test automation. It usually includes a test harness and test libraries.

Test basis The body of knowledge used as the basis for test analysis and design.

Test case A set of preconditions, inputs, actions (where applicable), expected results and postconditions, developed based on test conditions.

Test charter Documentation of the goal or objective for a test session. Refer also to: *Exploratory testing.* Synonym: charter

Test completion The activity that makes testware available for later use, leaves test environments in a satisfactory condition and communicates the results of testing to relevant stakeholders.

Test completion report A type of test report produced at completion milestones that provides an evaluation of the corresponding test items against exit criteria. Synonym: test summary report

Test condition A testable aspect of a component or system identified as a basis for testing. Synonyms: test situation, test requirement

Test control The activity that develops and applies corrective actions to get a test project on track when it deviates from what was planned. Refer also to: *Test management.*

Test cycle   An instance of the test process against a single identifiable version of the test object.

Test data   Data needed for test execution.
Synonym: test dataset

Test design   The activity that derives and specifies test cases from test conditions.

Test environment   An environment containing hardware, instrumentation, simulators, software tools and other support elements needed to conduct a test.
Synonyms: test bed, test rig

Test estimation   An approximation related to various aspects of testing.

Test execution   The activity that runs a test on a component or system producing actual results.

Test execution schedule   A schedule for the execution of test suites within a test cycle.

Test harness   A collection of stubs and drivers needed to execute a test suite

Test implementation   The activity that prepares the testware needed for test execution based on test analysis and design.

Test item   A part of a test object used in the test process. Refer also to: *Test object*.

Test level   A specific instantiation of a test process.
Synonym: test stage

Test log   A chronological record of relevant details about the execution of tests.
Synonyms: test record, test run log

Test management   The process of planning, scheduling, estimating, monitoring, reporting, controlling and completing test activities. Refer also to: *Test control, Test monitoring*.

Test manager   The person responsible for project management of testing activities, resources and evaluation of a test object.

Test monitoring   The activity that checks the status of testing activities, identifies any variances from planned or expected, and reports status to stakeholders. Refer also to: *Test management*

Test object   The work product to be tested. Refer also to: *Test item*

Test objective   The purpose for testing.
Synonym: test goal

Test plan   Documentation describing the test objectives to be achieved and the means and the schedule for achieving them, organized to coordinate testing activities.

Test planning   The activity of establishing or updating a test plan.

Test policy   A high-level document describing the principles, approach and major objectives of the organization regarding testing.
Synonym: organizational test policy

Test procedure   A sequence of test cases in execution order, and any associated actions that may be required to set up the initial preconditions and any wrap up activities post execution.

Test process   The set of interrelated activities comprising of test planning, test monitoring and control, test analysis, test design, test implementation, test execution and test completion.

Test progress report   A type of periodic test report that includes the progress of test activities against a baseline, risks, and alternatives requiring a decision.
Synonym: test status report

Test pyramid   A graphical model representing the relationship of the amount of testing per level, with more at the bottom than at the top.

Test report   Documentation summarizing test activities and results.

Test reporting   Collecting and analyzing data from testing activities and subsequently consolidating the data in a report to inform stakeholders.

Test result   The consequence/outcome of the execution of a test.
Synonyms: outcome, test outcome, result

Test run   The execution of a test suite on a specific version of the test object.

Test script   A sequence of instructions for the execution of a test.

Test session   An uninterrupted period of time spent in executing tests.

Test strategy   description of how to perform testing to reach test objectives under given circumstances.

Test suite   A set of test scripts or test procedures to be executed in a specific test run.
Synonyms: test set, test case suite

Test technique   A procedure used to define test conditions, design test cases, and specify test data.
Synonym: test design technique, test case design technique, test specification technique

Test type   A group of test activities based on specific test objectives aimed at specific characteristics of a component or system.

Test-driven development   A software development technique in which the test cases are developed, automated and then the software is developed incrementally to pass those test cases. Refer also to: *Test-first approach*.

Test-first approach   An approach to software development in which the test cases are designed and implemented before the associated component or system is developed. Refer also to: *Test-driven development*.

Testability   The degree to which test conditions can be established for a component or system, and tests can be performed to determine whether those test conditions have been met. Refer also to: *Maintainability*.

Tester   A person who performs testing.

Testing   The process within the software development lifecycle that evaluates the quality of a component or system and related work products. Refer also to: *Quality control*.

Testing quadrants   A classification model of test types/test levels in four quadrants, relating them to two dimensions of test objectives: supporting the product team versus critiquing the product, and technology-facing versus business-facing.

Testware   Work products produced during the test process for use in planning, designing, executing, evaluating and reporting on testing.

Traceability   The ability to establish explicit relationships between related work products or items within work products.

Unit test framework   A tool that provides an environment for unit or component testing in which a component can be tested in isolation or with suitable stubs and drivers. It also provides other support for the developer, such as debugging capabilities.

Usability   The degree to which a component or system can be used by specified users to achieve specified goals in a specified context of use.

Usability lab   Test facility in which unintrusive observation of participant reactions and responses to software takes place.

Usability testing   Testing to evaluate the degree to which the system can be used by specified users with effectiveness, efficiency and satisfaction in a specified context of use.

User acceptance testing   A type of acceptance testing performed to determine if intended users accept the system. Refer also to: *Acceptance testing*.

User experience   A person's perceptions and responses resulting from the use or anticipated use of a software product.

User story   A user or business requirement consisting of one sentence expressed in the everyday or business language which is capturing the functionality a user needs, the reason behind it, any non-functional criteria and also including acceptance criteria.

V-model   A sequential software development lifecycle model describing a one-for-one relationship between major phases of software development from business requirements specification to delivery, and corresponding test levels from acceptance testing to component testing.

Validation   Confirmation by examination that a work product matches a stakeholder's needs.

Verification   Confirmation by examination and through provision of objective evidence that specified requirements have been fulfilled.

Walkthrough   A type of review in which an author leads members of the review through a work product and the members ask questions and make comments about possible issues.
Synonym: structured walkthrough

White-box test technique   A test technique only based on the internal structure of a component or system.
Synonyms: white-box test design technique, structure-based test technique

White-box testing   Testing based on an analysis of the internal structure of the component or system.
Synonyms: clear-box testing, code-based testing, glass-box testing, logic-coverage testing, logic-driven testing, structural testing, structure-based testing

Wideband Delphi   An expert-based test estimation technique that aims at making an accurate estimation using the collective wisdom of the team members.

ANSWERS TO SAMPLE EXAM QUESTIONS

This section contains the answers and the learning objectives for the sample questions in each chapter and for the full mock exam in Chapter 7.

If you get any of the questions wrong or if you were not sure about the answer, then the learning objective tells you which part of the Syllabus to go back to in order to help you understand why the correct answer is the right one. The learning objectives are listed at the beginning of each section. For example, if you got Question 3 in Chapter 1 wrong, then go to Chapter 1 and read Learning Objective 1.2.2. Then re-read the section in the chapter which deals with that topic.

CHAPTER 1 FUNDAMENTALS OF TESTING

Question	Answer	Learning objective
1	a	1.1.1
2	a	1.1.2
3	c	1.2.2
4	a	1.1.1
5	a	1.3.1
6	c	1.4.4
7	b	1.5.2
8	d	1.4.3
9	b	1.5.3
10	d	1.4.5
11	b	1.4.5
12	a	1.4.3

CHAPTER 2 TESTING THROUGHOUT THE SOFTWARE DEVELOPMENT LIFE CYCLE

Question	Answer	Learning objective
1	d	2.1.1
2	c	2.1.3
3	b	2.2.2
4	b	2.3.1
5	c	2.2.2
6	d	2.3.1
7	c	2.2.3
8	b	2.2.2
9	c	2.1.5

CHAPTER 3 STATIC TESTING

Question	Answer	Learning objective
1	d	3.1.1
2	a	3.2.2
3	d	3.1.3
4	a	3.2.4
5	d	3.2.3
6	b	3.2.4
7	a	3.2.5
8	c	3.2.1
9	c	3.1.2

CHAPTER 4 TEST ANALYSIS AND DESIGN

Question	Answer	Learning objective
1	d	4.4.3
2	a	4.2.2
3	c	4.3.3
4	a	4.1.1
5	b	4.3.2
6	c	4.2.3
7	d	4.2.4
8	b	4.2.1
9	b	4.2
10	c	4.3
11	a	4.5.2
12	c	4.3
13	a	4.4
14	c	4.3.1
15	d	4.3
16	b	4.4.2
17	a	4.2.4
18	b	4.5.1
19	c	4.5.3

CHAPTER 5 **MANAGING THE TEST ACTIVITIES**

Question	Answer	Learning objective
1	c	5.1.4
2	d	5.1.3
3	a	5.1.1
4	c	5.3.1
5	b	5.3.2
6	a	5.5.1
7	c	5.4.1
8	d	5.2.2
9	b	5.2.1
10	a	5.2.2
11	a	5.2.4
12	b	5.1.4
13	d	5.5.1
14	c	5.1.5
15	a	5.1.3
16	c	5.1.2
17	a	5.1.3
18	d	5.1.4
19	d	5.1.5
20	d	5.1.6
21	d	5.1.7
22	c	5.2.4
23	b	5.3.3

CHAPTER 6 **TEST TOOLS**

Question	Answer	Learning objective
1	d	6.1.1
2	c	6.1.1
3	b	6.2.1
4	a	6.2.1

CHAPTER 7 MOCK EXAM

Question	Answer	Learning objective
1	a	1.1.2
2	c	1.4.2
3	c	1.4.1
4	d	1.1.1
5	d	1.5.2
6	d	1.5.3
7	c	1.3.1
8	c	1.4.5
9	b	2.2.3
10	b	2.2.2
11	a	2.1.2
12	b	2.1.3
13	c	2.1.5
14	b	2.1.6
15	b	3.1.1
16	c	3.1.3
17	b	3.2.5
18	b	3.2.4
19	b	4.2.1
20	b	4.1.1
21	a	4.3.2
22	c	4.4.1
23	a	4.2.1
24	c	4.2.2
25	c	4.2.3
26	d	4.4.2
27	c	4.2.4
28	b	4.5.1
29	b	4.3.3
30	c	5.2.2
31	b	5.5.1
32	b	5.3.1
33	c	5.1.1
34	d	5.4.1
35	c	5.1.5
36	b	5.1.7
37	d	5.3.3
38	b	5.1.4
39	b	6.1.1
40	a	6.2.1

REFERENCES

Key:
In Syllabus
Extra to Syllabus

CHAPTER 1 FUNDAMENTALS OF TESTING

Beizer, B. (1990) *Software Testing Techniques* (2nd edition), Van Nostrand Reinhold: Boston, MA

Black, R. (2004) *Critical Testing Processes*, Addison Wesley: Reading, MA

Black, R. (2009) *Managing the Testing Process* (3rd edition), John Wiley & Sons: New York, NY

Boehm, B. (1986) 'A Spiral Model of Software Development and Enhancement', *ACM SIGSOFT Software Engineering Notes*, ACM, 11(4):14–25, August 1986

Gilb, T. and Graham, D. (1993) *Software Inspection*, Addison Wesley: Reading, MA

NATS (2003) https://www.caa.co.uk/publication/download/20648

ISO/IEC/IEEE 29119-1 (2022) *Software and systems engineering – Software testing – Part 1: Concepts and definitions*

ISO/IEC/IEEE 29119-2 (2021) *Software and systems engineering – Software testing – Part 2: Test processes*

ISO/IEC/IEEE 29119-3 (2021) *Software and systems engineering – Software testing – Part 3: Test documentation*

Jones, C. (2008) *Estimating Software Costs* (3rd edition), McGraw Hill Education: New York, NY

Myers, G., Badgett, T. and Sandler, C. (2011) *The Art of Software Testing* (3rd edition), John Wiley & Sons: New York, NY

Weinberg, G. (2008) *Perfect Software and Other Illusions about Testing*, Dorset House: New York, NY

CHAPTER 2 TESTING THROUGHOUT THE SOFTWARE DEVELOPMENT LIFE CYCLE

Berard, E. V. (1993) *Essays on Object-oriented Software Engineering* (Volume 1), Prentice Hall: Englewood Cliffs, NJ

Black, R. (2017) *Agile Testing Foundations*, BCS Learning & Development Ltd: Swindon, UK

Boehm, B. (1986) 'A Spiral Model of Software Development and Enhancement', *ACM SIGSOFT Software Engineering Notes*, ACM, 11(4):14–25, August 1986

Crispin, L. and Gregory, J. (2008) *Agile Testing*, Pearson Education: Boston, MA

Gregory, J. and Crispin, L. (2015) *More Agile Testing*, Pearson Education: Boston, MA

ISO/IEC 25010 (2011) *Systems and software engineering – Systems and Software Quality Requirements and Evaluation (SQuaRE) System and Software Quality Models*

CHAPTER 3 STATIC TECHNIQUES

Gilb, T. and Graham, D. (1993) *Software Inspection*, Addison-Wesley: London

ISO/IEC 20246 (2017) *Software and system engineering – Work product reviews*

Kramer, A. and Legeard, B. (2016) *Model-Based Testing Essentials: Guide to the ISTQB Certified Model-Based Tester: Foundation Level*, John Wiley & Sons: New York, NY

NASA (1999) https://en.wikipedia.org/wiki/Mars_Climate_Orbiter

Sauer, C. (2000) 'The Effectiveness of Software Development Technical Reviews: A Behaviorally Motivated Program of Research', *IEEE Transactions on Software Engineering*, 26(1):1–14

Shull, F., Rus, I. and Basili, V. (2000), 'How Perspective-Based Reading Can Improve Requirement Inspections', *IEEE Computer*, 33(7):73–79

van Veenendaal, E. (1999) 'Practical Quality Assurance for Embedded Software', in *Software Quality Professional*, Vol. 1, no. 3, American Society for Quality, June 1999

van Veenendaal, E. (ed.) (2004) *The Testing Practitioner* (Chapters 8–10), UTN Publisher: The Netherlands

van Veenendaal, E. and van der Zwan, M. (2000) 'GQM Based Inspections', in *Proceedings of the 11th European Software Control and Metrics Conference (ESCOM)*, Munich, May 2000

Wiegers, K. (2002) *Peer Reviews in Software*, Pearson Education: Boston, MA

CHAPTER 4 **TEST TECHNIQUES**

Adzic, G. (2009) *Bridging the Communication Gap: Specification by Example and Agile Acceptance Testing*, Neuri Limited

Ammann, P. and Offutt, J. (2016) *Introduction to Software Testing* (2nd edition), Cambridge University Press: Cambridge, UK

Andrews, M. and Whittaker, J. (2006) *How to Break Web Software: Functional and Security Testing of Web Applications and Web Services*, Addison-Wesley Professional: Reading, MA

Beizer, B. (1990) *Software Testing Techniques* (2nd edition), Van Nostrand Reinhold: Boston, MA

Black, R. (2007) *Pragmatic Software Testing*, John Wiley & Sons: New York, NY

Broekman, B. and Notenboom, E. (2003) *Testing Embedded Software*, Addison Wesley: London

Brykczynski, B. (1999) 'A Survey of Software Inspection Checklists', *ACM SIGSOFT Software Engineering Notes*, 24(1), pp. 82–89

Cohn, M. (2009) *Succeeding with Agile: Software Development Using Scrum*, Addison-Wesley: Reading, MA

Copeland, L. (2004) *A Practitioner's Guide to Software Test Design*, Artech House: Norwood, MA

Craig, R. D. and Jaskiel, S. P. (2002) *Systematic Software Testing*, Artech House: Norwood, MA

Forgács, I. and Kovács, A. (2019) *Practical Test Design: Selection of Traditional and Automated Test Design Techniques*, BCS, The Chartered Institute for IT

Gilb, T. (1988) *Principles of Software Engineering Management*, Addison Wesley: Reading, MA

Hendrickson, E. (2013) *Explore It!: Reduce Risk and Increase Confidence with Exploratory Testing*, The Pragmatic Bookshelf: Dallas TX

Hetzel, B. (1988) *The Complete Guide to Software Testing* (2nd edition), QED Information Sciences: Wellesley, MA

ISO/IEC/IEEE 29119-4 (2021) *Software and System Engineering - Software Testing - Part 4: Test Techniques*

Jeffries, R., Anderson, A. and Hendrickson, C. (2000) *Extreme Programming Installed*, Addison-Wesley Professional: Reading, UK

Jorgensen, P. (2014) *Software Testing: A Craftsman's Approach* (4th edition), CRC Press: Boca Raton, FL

Kaner, C., Falk, J. and Nguyen, H. Q. (1999) *Testing Computer Software* (2nd edition), John Wiley & Sons: New York, NY

Kaner, C., Bach, J. and Petticord, B. (2002) *Lessons Learned in Software Testing*, John Wiley & Sons: New York, NY

Kaner, C., Padmanabhan, S. and Hoffman, D. (2013) *The Domain Testing Workbook*, Context-driven Press: Orlando, FL

Koomen, T., van der Aalst, L., Broekman, B. and Vroon, M. (2006) *TMap Next for Result-driven Testing*, UTN Publishers, The Netherlands

Marick, B. (1994) *The Craft of Software Testing*, Prentice Hall: New York, NY

Myers, G., Badgett, T. and Sandler, C. (2011) *The Art of Software Testing* (3rd edition), John Wiley & Sons: New York, NY

Nielsen, J. (1994) 'Enhancing the Explanatory Power of Usability Heuristics', *Proceedings of the SIGCHI Conference on Human Factors in Computing Systems: Celebrating Interdependence*, ACM Press, pp. 152–158

O'Regan, G. (2019) *Concise Guide to Software Testing*, Springer Nature: Cham, Switzerland

Pol, M., Teunissen, R. and van Veenendaal, E. (2001) *Software Testing: A Guide to the TMap Approach*, Addison Wesley: Harlow, UK

Sabourin, R. (2024), *Charting the Course: Coming up with Great Test Ideas; Just in Time*, Notion Press: India, Singapore, Malaysia.

Watson, A. H., Wallace, D. R. and McCabe, T. J. (1996) *Structured Testing: A Testing Methodology Using the Cyclomatic Complexity Metric*, U.S. Dept. of Commerce, Technology Administration, NIST

Whittaker, J. A. (2003) *How to Break Software: A Practical Guide to Testing*, Addison Wesley: Reading, MA

Whittaker, J. (2009) *Exploratory Software Testing: Tips, Tricks, Tours, and Techniques to Guide Test Design*, Addison Wesley: Reading, MA

CHAPTER 5 **TEST MANAGEMENT**

Beizer, B. (1990) *Software Testing Techniques* (2nd edition), Van Nostrand Reinhold: Boston, MA

Black, R. (2004) *Critical Testing Processes*, Addison Wesley: Reading, MA

Black, R. (2009) *Managing the Testing Process* (3rd edition), John Wiley & Sons: New York, NY

Brooks, F. (1995) *The Mythical Man-Month and Other Essays on Software Engineering*, Addison Wesley: New York, NY

Hetzel, W. (1988) *Complete Guide to Software Testing*, QED: Wellesley, MA

ISO/IEC/IEEE 29119-3 (2013) *Software and Systems Engineering – Software Testing – Part 3: Test documentation*

ISO/IEC 25010 (2011) *Systems and Software Engineering – Systems and Software Quality Requirements and Evaluation (SQuaRE) System and Software Quality Models*

Pol, M., Teunissen, R. and van Veenendaal, E. (2002) *Software Testing: A Guide to the TMap Approach*, Addison Wesley: Reading, MA

van Veenendaal, E. and Wells, B. (2012) *Test Maturity Model Integration TMMi: (Guidelines for Test Process Improvement)*, UTN Publishers: Den Bosch, The Netherlands

CHAPTER 6 **TEST TOOLS**

Axelrod, A. (2018) *Complete Guide to Test Automation: Techniques, Practices and Patterns for Building and Maintaining Effective Software Projects*, apress: New York, NY

Buwalda, H., Janssen, D. and Pinkster, I. (2001) *Integrated Test Design and Automation*, Addison Wesley: Reading, MA

Fewster, M. and Graham, D. (1999) *Software Test Automation*, Addison Wesley: Reading, MA

Gamba, S. and Graham, D. (2018) *A Journey Through Test Automation Patterns*, amazon

Graham, D. and Fewster, M. (2012) *Experiences of Test Automation*, Pearson Education: Boston, MA

Graham, D. and Gamba, S. (2018) https://TestAutomationPatterns.org

AUTHORS

ERIK VAN VEENENDAAL

Erik van Veenendaal is an internationally recognized testing expert, author of a number of books and has published a large number of papers within the profession. He is currently working as an independent consultant and as the Chief Executive Officer (CEO) of the TMMi Foundation.

Dr Erik van Veenendaal CISA graduated from the University of Tilburg in Business Economics. He has been working as a practitioner and manager in the IT industry since 1987. After a career in software development, he moved to the area of software quality, where he specializes in software testing.

As a test manager and test consultant he has been involved in a great number and variety of projects. He has implemented structured testing and reviews and inspections, and as a consultant has contributed to many test process improvement projects. He worked for Sogeti as manager of operations and was one the core developers of the TMap testing methodology. He is the author of numerous papers and a number of books on testing and software, including the best sellers *The Testing Practitioner, The Little TMMi, Foundations of Software Testing* and *Testing according to TMap*.

Erik van Veenendaal founded Improve Quality Services BV, a company that provides consultancy and training services in the areas of testing, requirements engineering and quality management. He was the company director for over 12 years. Within this period Improve Quality Services became a leading testing company in The Netherlands focused on innovative and high-quality testing services. Customers were especially to be found in the areas of embedded software (e.g. Philips, Océ en Assembléon) and in the finance domain (e.g. Rabobank, ING and Triodos Bank). Improve Quality Services was market leader for test training in The Netherlands both in terms of quantity and quality.

Nowadays he is occupied as an independent consultant doing consultancy and training in the areas of testing (especially based on ISTQB Syllabi) and requirements engineering. He also publishes and delivers keynote presentations on a regular basis. In addition, he is actively involved in the TMMi Foundation, International Software Testing Qualifications Board and the IREB requirements engineering organization.

Erik, being one the initiators who founded the TESTNET organization, is now an honorary member of TESTNET (the Dutch Special Interest Group in Software Testing). Erik was the first person to receive the Information Systems Examination Board (ISEB) Practitioner certificate with distinction and is also a Certified Information Systems Auditor (CISA). Erik was also a senior lecturer at the Eindhoven University of Technology, Faculty of Technology Management from 1996 until 2005.

Since its foundation in 2002, Erik has been strongly involved in the International Software Testing Qualifications Board (ISTQB). From 2005 till 2009 he was the vice-president of the ISTQB organization and he is the founder of the local Belgium and The Netherlands board; the Belgium Netherlands Testing Qualifications Board (BNTQB), and the Curaçao board; the Curaçao Testing Qualifications Board (CTQB). For many years, he was the editor of the ISTQB *Standard Glossary of Terms used in Software Testing* and chair of the ISTQB Expert level working party. Today, he is a CTQB board member. For his major contribution to the field of testing, Erik received the European Testing Excellence Award in December 2007 and the ISTQB International Testing Excellence Award in October 2015.

Erik is blessed with three great kids (Tim, Lars and Anne). In addition to family and work, Erik likes to spend time on sports. He has been playing table tennis since his youth and played at the world championships in Rotterdam (The Netherlands, 2011) and Dortmund (Germany, 2012). During the Caribbean championships in Havana (Cuba) in 2014 he won the bronze medal. He also plays the guitar and is a huge Bruce Springsteen fan and has been to several of his concerts throughout Europe.

Erik can be contacted via email at erik@erikvanveenendaal.nl and through his website www.erikvanveenendaal.nl.

REX BLACK

Rex Black started in software and systems engineering in 1983, writing FORTRAN and C code for financial applications. In the late 1980s, he transitioned into software and system testing, focusing on test automation and test management. In the mid 1990s, he started a consulting company, RBCS (now Rex Black, Inc.), which provides consulting and training services related to software testing, software quality, software development, software project management and DevOps.

Including this book, Rex has written extensively on these topics, including *Managing the Testing Process, Advanced Software Testing: Volume I, Advanced Software Testing: Volume II, Advanced Software Testing: Volume III, Critical Testing Processes, Pragmatic Software Testing, Agile Testing Foundations, Mobile Testing, Expert Test Manager, AI and Software Testing,* and *Fundamentos de Prueba de Software.*

These books have sold hundreds of thousands of copies, including Hebrew, Indian, Chinese, Hungarian, Japanese and Russian editions. He has written over 30 articles; presented hundreds of papers, workshops and seminars; and given about 50 keynote and other speeches at conferences and events around the world.

Rex also helped found the ISTQB, served as President for four years, and contributed to most of the intellectual property developed between 2003 and 2018. In addition, Rex helped found the ASTQB and TMMi North America. Now, Rex works as a QA Architect for Epic Games, applying 40 years of experience to help build the most amazing software testing team of his career in one of the most challenging software testing and quality domains.

Rex is married to Laurel Becker. They met in 1987 at the University of California, Los Angeles. They have two children, Emma and Charlotte, and three dogs, Kibo, Mmink and Roscoe. They live in Bulverde, Texas and Lake Tahoe, California.

DOROTHY GRAHAM

Dorothy Graham (Dot) has been involved in software testing since 1970, when her first job (for Bell Labs in the US) was as a programmer in a testing group, and her task was to write two testing tools (test execution and comparison). After emigrating to the UK (with her British husband Roger), she worked for Ferranti Computer Systems, developing software for UK Police forces. She then joined the National Computing Centre as a trainer and courseware developer, and later became an independent consultant.

At that time, software testing was not a respected profession; in fact, in the early 1990s, many thought of testing at best as a necessary evil (if they thought of testing at all!). There were few people who specialized in testing, and it was seen as a second-class activity and not well thought of. There was a general perception that testing was easy, that anyone could do it, and that you were rather strange if you liked it. It was then that Dot decided to specialize in testing, seeing great scope for improvement in testing activities in industry, not only in imparting fundamental knowledge about testing (basic principles and techniques) but also in improving the view testers had of themselves, and the perceptions of testers in their companies. She developed training courses in testing and began Grove Consultants, named after her house in Macclesfield, UK. One of her most popular talks at the time was called *Test is a four-letter word*, reflecting the prevailing culture about testing.

It was into this context that the initiative to create a qualification for testers was born. Although not the initiator, Dot was involved from the first meetings and the earliest working groups that developed the first Foundation Syllabus, donating many hours of time to help progress this effort. This work was carried out with support from ISEB (Information Systems Examination Board) of the British Computer Society, and the testing qualification was modelled on ISEB's successful qualifications in Project Management and Information Systems Infrastructure. One of the aims at this time was to give people a common vocabulary to talk about testing, since people seemed to be using many different terms for the same thing.

Grove Consultants (Dorothy Graham and Mark Fewster at that time) gave the first course based on the ISEB Foundation Syllabus in October 1998 and the first Foundation Certificates in Software Testing were awarded. In her 20 years with them, Grove went on to be highly respected for the quality of its training material, particularly for ISTQB courses. Grove (now Grove Software Testing Ltd) continues to license high-quality ISTQB training material to organizations wishing to provide training without expending a significant effort in course development (https://grove.co.uk). Dot is still involved as adviser to Grove.

The success of the Foundation qualification took everyone by surprise. There seemed to be a hunger for a qualification that gave testers more respect, both for themselves and from their employers. It also gave testers a common vocabulary and more confidence in their work. The Foundation qualification had met its main objective of 'removing the bottom layer of ignorance' about software testing.

Work then began on extending the ISEB qualification to a more advanced level (which became the ISEB Practitioner qualification) and also to extending it to other countries, as news of the qualification spread in the international community. Dot was a facilitator at the meeting that formed ISTQB in 2001 in Sollentuna, Sweden. She has not been actively involved in the ISTQB Advanced levels. She helped to write the first ISTQB Foundation Syllabus (with Erik and others).

Dorothy's other activities over the years include being Programme Chair for the first European testing conference (EuroSTAR 1993); she was Programme Chair again in 2009. During the 1990s, she started and later co-authored *The CAST Report*, a summary of commercial testing tools (in the days before the internet!). In addition to all editions of this book on Foundations of Software Testing, she is co-author of four other books: *Software Inspection* (1993) with Tom Gilb, *Software Test Automation* (1998) and *Experiences of Software Test Automation* (2012), both with Mark Fewster, and *A Journey Through Test Automation Patterns* (2018) with Seretta Gamba.

Her book with Seretta is the story of a team using the test automation patterns wiki, https://TestAutomationPatterns.org. This wiki, developed by Seretta and Dot, provides solutions that have worked for others for a number of issues and problems in test automation. The issues and patterns are organized into four sections: Process, Management, Design and Execution. The wiki was first published in 2013 and is a popular source of advice about test automation.

During her career, Dot has spoken at over 400 testing conferences and events worldwide, including over 200 major conferences. She was awarded the second European Excellence Award in Software Testing in 1999 and was awarded the first ISTQB Excellence Award in Software Testing in 2012.

Her main non-testing activities include singing (choirs, madrigal groups and solos) and enjoyable cruises and holidays with her husband, Roger. They have two children, Sarah (married to Tim) and James.

Now mostly retired, Dorothy can be contacted on LinkedIn and info@DorothyGraham.co.uk.

INDEX